THE PHYSICAL LANDSCAPE IN PICTURES

by A. V. HARDY AND F. J. MONKHOUSE

I	The Earth	*page* 1
II	Earth Movements	13
III	Vulcanicity	20
IV	Weathering and Soils	29
V	Limestone and Chalk Country	35
VI	Rivers	39
VII	Glaciation	52
VIII	Lakes	60
IX	Deserts	64
X	Coastlines	70
XI	Vegetation	85

CAMBRIDGE · AT THE UNIVERSITY PRESS · 1964

PUBLISHED BY

THE SYNDICS OF THE CAMBRIDGE UNIVERSITY PRESS

Bentley House, 200 Euston Road, London, N.W.1

American Branch: 32 East 57th Street, New York 22, N.Y.

West African Office: P.O. Box 33, Ibadan, Nigeria

Printed in Great Britain by Jarrold & Sons Ltd, Norwich

Preface

Geographers differ considerably in their interpretation of the subject, but most would agree that one of the main aims of teaching is to increase the student's knowledge of the world. To observe and to explain the major features of the landscapes in all their variety which the world offers are primary objects. In performing this task, a photograph ranks as one of the main sources of material because, especially when used in conjunction with a map, it is the next best thing to actual study in the field. In one important sense, a photograph needs no interpretation because an immediate vivid impression of the whole of the visible scene is gained by all who will but see. Yet, in another sense, the ability to use a photograph to the best advantage increases with our knowledge and experience as geographers. It needs to be studied, appreciated, interpreted and understood.

This book is an attempt to illustrate the main features of the 'natural' or 'physical' landscape by means of a carefully chosen series of photographs. The text has deliberately been kept as short and elementary as possible, and is intended mainly to introduce, comment upon, supplement and where necessary explain the illustrations. Nevertheless, because these have been selected to cover as widely as possible the recognised course of work, the text also is complete in itself. Each section of the book has as its theme a distinctive element in the study of the physical landscape and is related to what comes both before and after. An attempt has been made to illustrate and define in passing the commoner technical terms used in current textbooks.

The examples have been selected to arouse the interest and meet the needs of Ordinary Level and School Certificate students in all types of secondary schools at home and overseas; the text has similarly been written at this level. It is hoped, however, that teachers will make use of this illustrative material in other parts of the school. With a few exceptions the photographs have not previously been published, and a number have been taken specifically for this book; one or two have already been set for the G.C.E. examination.

F.J.M.
A.V.H.

ACKNOWLEDGEMENTS

Our thanks for permission to reproduce photographs are due to the following people and organisations:

Aerofilms Ltd for 3, 24, 45, 47, 55, 58, 64, 71 and 72; Associated Press for 1; J. Allan Cash for 75; M. J. Chambers for 57; *Coal News* for 8; Ewing Galloway, N.Y., for 37; Fairchild Aerial Surveys, Inc., for 9, 20, 41 and 54; *Geological Survey* for 4, 5, 6, 38 and 49; R. Kay Gresswell for 14; Hunting Aerosurveys Ltd for 12; Eric Kay for 2, 7, 10, 11, 17, 18, 19, 27, 29, 30, 31, 32, 43, 48, 50, 59, 62 and 63; L. J. Monkhouse for 16, 23, 34, 36 and 56; Photographic Survey Corporation for 13, 39 and 74; Paul Popper for 69, 70 and 73; J. K. St Joseph for 42, 53, 60 and 65; Swissair for 44 and 46; the U.S. Air Force for 40; and White's Aviation Ltd for 21, 66, 67 and 68. Photographs 15, 22, 25, 26, 28, 33, 35, 51, 52 and 61 are by F. J. Monkhouse.

I. The Earth

1 The remarkable photograph on this page was taken by astronaut Colonel John H. Glenn as his 'Mercury' capsule orbited the earth on 20 February 1962, at a speed of over 17,000 miles an hour. The picture shows the earth's surface and atmosphere lit up by the sun, though in part covered with scattered cloud which appears white from above. At the top of the picture, the curvature of the earth is clearly visible, with the void of outer space in black.

Events and pictures such as this demonstrate what has long been known, that the earth is a ball rotating in space. It is probable that it was a very different shape when it first came into existence. It has been gradually modified until it is now a spheroid (i.e. 'almost a sphere'), wherein almost every part of the earth's surface lies at a similar distance from the centre, and hence from its centre of gravity. The slight flattening of the earth's surface towards the Poles is probably the result of the rotation of the earth in space.

The conditions which permit life, as we know it, to exist on the earth are numerous, complicated and interrelated. Distance from

1 Astronaut's view of the earth

the sun, together with the depth and composition of the thin surrounding belt of atmosphere, jointly determine the amount of heat available to warm the earth's surface. It is one of the most remarkable facts of the universe that the earth's surface experiences a range of temperature which permits life to exist, for this would not be possible if the quantity of heat were appreciably greater or less. Sir James Jeans pointed out many years ago that the infinity of outer space, relative to which our earth is but a grain of sand, is extremely cold, except for the immediate neighbourhood of the vast number of stars, within each of which the temperature attains thousands of degrees Centigrade. Life can therefore only exist in a narrow temperate zone which surrounds each of these 'celestial fires' at a very definite distance, but which together constitute only an extremely small proportion of the whole of space. It has also been estimated that only one star in 100,000 has a planet revolving round it within the zone in which life is possible.

One of the essential conditions for life is a supply of water, the circulation of which in its various forms is a complicated process. The water vapour in the atmosphere coalesces into water droplets, which fall in frozen crystalline form as snow to nurture the snow-fields, ice-caps and glaciers of the Arctic and Antarctic regions and the high mountain ranges, or as rain to fill the rivers and lakes, and ultimately the seas and oceans. Much of this water is once again absorbed by evaporation into the atmosphere from the earth's surface, and hence the circulation is completed. It should be noted that no supply of water is available from outside the earth and its atmosphere. Only because of this circulation of water is the earth able to retain its characteristic distribution of oceans, seas and rivers.

The earth probably originated as a gaseous mass, which gradually cooled, so that the surface crust has long since solidified. The composition of the interior is not known except in theory, but it is estimated from the calculated density of the earth as a whole that it must be composed of very dense material, possibly in a molten state, though under intense pressure. It may well be that the continents are 'floating' like rafts on a molten foundation.

The earth's crust is not completely quiet or stable; earthquakes, volcanoes and other crustal activity show that, beneath the usually placid surface, changes are still in progress. From time to time part of the underlying molten rock material has approached or actually reached the surface. Such material, on cooling and solidifying, forms much of the crust, and is given the name of *igneous* (that is, 'fire-formed') rock. Granite and basalt are two of the most common igneous rocks.

Once a solid crust of some kind had formed on the earth, weathering and erosion began to wear away the rocks. Fragments were deposited in or on the margins of lakes, seas and oceans, as well as on the land itself. Over long periods of time, layer upon layer of material was deposited in this way to form what are called the *sedimentary* rocks; clay, sandstone, chalk and limestone are some of the most widespread.

In addition to these igneous and sedimentary rocks, a third class is named *metamorphic* rock. These are rocks, of igneous or sedimentary origin, which have been greatly changed in character by heat or pressure, or both. Very hot material from the interior of the earth, forcing its way upwards through layers of other rocks, has often completely altered their structure and composition. During earth movements rocks are frequently subject to intense pressure, and therefore are altered in character. For example, limestone has been changed into marble, shale into slate and schist, sandstone into quartzite, and granite into gneiss.

2 Haytor, Dartmoor, Devon

2 The photograph above is of Haytor in Devonshire, one of the many 'tors' which are to be found on Dartmoor. It will be seen from the figures standing on the highest part that Haytor is of considerable size. These tors are masses of granite which have been left upstanding whilst the surrounding land has been worn away to a lower level.

A close examination of a piece of granite will show that it is composed of a mass of crystals of different sizes and colours. The most numerous of these, small and light in colour, are called felspar. Specks of mica give a sparkle to the rock, and other grey or white crystals made of quartz form a third easily identifiable substance. Felspar, mica and quartz are known as *minerals*. The materials of the earth's crust (the *rocks*) contain a large number of minerals, though the important ones, from a geographer's point of view, are few in number. Limestone and chalk are mainly composed of calcite (calcium carbonate); basalt contains much augite—composed of fine black crystals—as well as felspars. Sandstones are largely formed of quartz, and shales of finely crushed felspar, mica and clay minerals.

Quartz, like most minerals, has a definite chemical composition—SiO_2 or silica—and is formed of the two common elements oxygen and silicon, in the proportion of 2:1. In fact, a very small number of chemical elements—oxygen, silicon, aluminium, iron, calcium, sodium, potassium, magnesium and hydrogen—make up about 98 per cent of all rocks. Most minerals, like quartz, consist of two or more elements, but a few, such as gold, tin and copper, form minerals by themselves.

Most of us have at one time or another produced crystals in the chemical laboratory by heating a strong solution of sugar or some other substance. Under favourable conditions many minerals produce crystals, and hence we refer to one group of rocks as crystalline. In a crystalline structure, the atoms are arranged in patterns which give the crystals both their characteristic beauty and frequently great strength and hardness. Quartz crystals, for example, are too hard to be scratched with a knife. Mainly for this reason crystalline rocks such as granite are usually harder than non-crystalline rocks.

Granite is found extensively in the mountainous parts of the British Isles, in Dartmoor and Bodmin Moor in the south-west, in North Wales and the English Lake District, and in many parts of Scotland, notably in the Cairngorms, the Isle of Arran and south-eastern Skye.

One other feature clearly visible in the Haytor rocks is the system of *joints* or cracks, both vertical and horizontal. These are set up in igneous rocks when the molten material cools and solidifies. They may also be formed by stresses caused by earth movements (see Fig. 14), and in sedimentary rocks during their consolidation. These cracks help to shape the form of the rock-masses. In some sandstones and limestones the joints cut across the near-horizontal strata, so dividing the rocks into rectangular blocks, known to quarrymen as *freestones*. These obviously are convenient for building, since the blocks are easy to extract and need relatively little cutting.

3 The Giant's Causeway, Antrim, Northern Ireland

3 The Giant's Causeway, shown opposite, is part of a series of cliffs on the coast of Northern Ireland. It is formed of an igneous rock called basalt, a heavy material possibly similar in type to the material which lies beneath the surface rocks of the earth. The crystals in granite and similar rocks, which cooled slowly beneath the surface, are large enough to be seen by the naked eye, but basalt appears to be a black featureless rock, which only the microscope reveals as a mass of tiny crystals, formed by the more rapid cooling at or very near the surface. In the Giant's Causeway, the contraction of the basalt in cooling has produced a series of prismatic columns, mainly hexagonal in form. The photograph shows that, despite the hardness of the rock, weathering and erosion by the sea have worn away great gullies in the cliffs.

Basalt, like granite, is found in the mountainous northern and north-western parts of the British Isles. Fingal's Cave, on the island of Staffa in western Scotland, has been made famous by Mendelssohn, and large parts of the Inner Hebrides, notably Skye and Eigg, consist of basalt. The rock often forms great plateaux or terraced hills, as in the north-western Deccan in India (which covers about 250,000 square miles) and the Columbia–Snake Plateau of the U.S.A. (see Fig. 22).

4 It may seem strange that sandstone should be composed mainly of particles of quartz similar to the fine grains of the sea-shore. The explanation is that the grains in sandstone are 'cemented' together by a material such as calcium carbonate, or by various iron compounds, or even, as in the quarry in the photograph below, by the same substance as the grains, silica. The grains of a sandstone which is soft and crumbly are usually loosely cemented together, while the grains on the sea-shore are not cemented at all.

4 Sandstone quarry in the eastern High Weald, Sussex

Three distinct layers (or *strata*) can be seen in the photograph, taken near Northiam in eastern Sussex. Underneath the clayey subsoil with small pebbles at the top of the photograph lies a layer of a hard buff-coloured sandstone. Below this (at the level of the hammer) is a layer of sandstone with a high iron content, in thinner bands. Below this is the main layer of solid, massive sandstone, 3 feet in thickness, the chief source of building stone in this neighbourhood, since it is easily cut out in uniform blocks. These rocks outcrop to the north of Hastings, and so are called by geologists the Hastings Beds. Softer sandstones are valuable as a source of building sand for making concrete and mortar. In the west and the Midlands of England, the New Red Sandstone—so coloured by its high iron content—forms farmland with deep red soils. By comparison, in the Pennines of northern England and in South Wales the Millstone Grit (see Fig. 27), in which the sand grains are unusually large and coarse and include numerous small pebbles, forms much bleak moorland country.

The photograph shows clearly that the sandstone is found in layers, and reminds us that sandstones are sedimentary rocks, formed from the debris of other rocks. This was transported by rivers to the sea, or was blown by the wind into dunes, where it accumulated in layers, to be consolidated by the pressure of later overlying deposits.

5 This photograph is of limestone composed almost entirely of shells; their size can be estimated in relation to the penny in the left of the picture. The calcite (calcium carbonate), of which limestone and chalk are composed, was formed from the remains of millions of creatures which lived in bygone ages in shallow, warm seas. It is still being formed in the coral

5 Shelly limestone, Viewedge Works, near Stokesay, Shropshire

reefs of the tropical parts of the world, and in fact in any waters where shells can accumulate.

Limestones vary greatly in colour, in hardness and in structure. The Oolitic Limestone of the Cotswold Hills in the west of England is formed of rounded grains, like fish-roe, hence the name, whereas the harder Carboniferous Limestone produces sizeable blocks of stone. While chalk is mainly white in colour (Figs. 32, 59), limestones may be white, grey, yellow or brown, varying in hardness and resistance to denudation. The scarplands of southern and south-eastern England are largely composed of chalk and limestone; Carboniferous Limestone forms much of the Pennines (Fig. 31) and underlies most of central Ireland. Some varieties afford excellent building stone, such as Portland Stone, Ham Stone, Bath Stone and 'Purbeck Marble', most of which occur as freestones (see p. 4). The rock which is too broken, soft or otherwise unsuitable for building purposes can be crushed for making lime and cement.

6 Fossils in Lias Shale, Antrim, Northern Ireland

6 The age of the earth cannot be exactly determined, but the oldest known rocks, which have been shown to exceed 3000 million years, are the hard, crystalline, metamorphic varieties, such as gneiss, schist and slate, of the Canadian Shield. The very old rocks form a large part of the earth's crust, and are of importance to man because they contain nearly all the world's deposits of certain metallic minerals such as gold, silver, uranium and copper. The coal-bearing rocks are much younger, and those containing oil are younger still.

Some of the life on the earth which existed during the millions of years in which the rocks were formed has been preserved in the form of *fossils*—the remains of animals and plants embedded in the rocks. They may be either the preserved hard parts (skeletons), or their exact replacement by some substitute such as silica or calcium carbonate. The oldest rocks reveal few fossils, but later ones contain the remains of different types of life, ranging from the bones of Dinosaurs, huge reptiles 14 feet high which lived about 140 million years ago, in Jurassic times, to the tiny shells of Foraminifera, one-celled sea-creatures whose remains are only visible through a microscope. Similarly, fossil plants vary from whole tree trunks converted into stone (as in the famous Petrified Forest of Arizona) to tiny organisms known as diatoms. In addition to actual fossils, the rocks contain large numbers of moulds and casts made by animals and plants such as worms and seaweed, and even footprints of long extinct animals. Our knowledge of the life of bygone ages is very incomplete, partly because only the hard parts of animals have survived and partly because much of the evidence is buried far below the surface and beneath the seas and oceans.

William Smith discovered over a century ago that because fossils varied considerably from one rock layer to another, it was possible to correlate and classify the rocks by age according to the fossils contained in them, as well as by the order in which the various layers of rock were laid down. On this basis, geologists have now divided the earth's history into five great *eras*, subdivided into *periods*, then into *epochs*, and finally into *ages*. The main eras are named from compounds of the Greek word *zoe*, meaning life. The earliest, which lasted from the earth's creation until about 600 million years ago, are known as *Azoic* ('without life') and *Proterozoic* ('early life'). The ancient rocks of these eras are usually grouped together as Pre-Cambrian or Archaean. Then come the *Palaeozoic* ('ancient life'), *Mesozoic* ('middle life'), and *Cainozoic* ('young life') eras. These three are sometimes referred to respectively as Primary, Secondary and Tertiary, while a fourth division (the Quaternary) was later added to represent both the time-duration and the rocks of the last 600,000 years or so.

As geographers we are not specially concerned with these details of the earth's geological history, but we are interested in the general characteristics of the rocks, both because of their usefulness to man and also because they help to determine the landscape. The old hard crystalline rocks of the Pre-Cambrian and Palaeozoic eras are able to support only a small fraction of the world's population. The great majority of people are found in the areas of younger rocks. In the British Isles, the old hard rocks occur in the north and west, comprising the highlands which are areas of scanty population; the south and east are formed of younger rocks in the shape of low hills, valleys and plains, most of which are densely peopled.

7 Opposite is a photograph of china-clay pits in Cornwall in south-western England. From pits such as these pure white clay is shipped in large quantities to the Potteries in the English Midlands and elsewhere for the manufacture of china and porcelain. Various kinds of clay have been used by Man for thousands of years for the production of all kinds of vessels. Clay differs as a sedimentary rock from sandstone because it is of very fine-grained material, containing much less quartz and a larger proportion of felspar and mica. It may be of many different colours. A very important use of clay is for the making of bricks and tiles. When wet, the minute pore spaces between the clay particles are filled with water, and so seal the rock against the downward passage of water (i.e. the clay is *impermeable*).

7 China-clay pit, near Penwithick, Cornwall

'Puddled clay' is sometimes used to line ponds and small reservoirs.

Whereas china-clay is formed from the decomposition of granite and is of unusual occurrence, many other types of clay are widespread. Boulder-clay, for example, is found on the surface over a large part of the Northern Hemisphere, where it was deposited during the Ice Age (see Fig. 50). Quite different is the London Clay which underlies much of the lower Thames Basin; this is a stratified sedimentary rock of considerable thickness.

Clay of one kind or another forms the floors of some of the largest of the world's lowlands, and provides the basis of much of the soil.

8 The most remarkable fossil plants are the remains of forests which helped to form the Coal Measure rocks. Coal was created by long processes which began when great forests of past ages were gradually covered with sediments brought down by rivers. After millions of years this material has been turned into coal. Similar processes are still at work in many parts of the world where shallow lakes are being filled with decaying vegetation to form bogs. This vegetation is ultimately converted into peat—a brown fibrous material, providing a slow-burning fuel still used in parts of Ireland and Scotland. Peat, when further compacted by the pressure of layers of sand and clay deposited above, slowly changes into 'brown

8 Underground coal-mining at the Wernos Colliery, Carmarthen, South Wales

coal'. Though of low quality, this is worked extensively in western Europe and the U.S.S.R. and is mainly used for the production of thermal electricity. The quality of coal proper depends on the proportion of carbon and volatile matter contained within it. The bituminous coals mined in Great Britain are of high quality. Anthracite contains some 93 to 95 per cent carbon, and is used largely for central heating.

Coal is found in thin seams separated from each other by layers of shale and sandstones, which together comprise the Coal Measures. A great belt of these rocks, which extends with interruptions through western Europe to the Donetz field of the U.S.S.R. in the Ukraine, was laid down in a series of basins. A large part of these rocks has been worn away, and in those that remain much of the coal near the surface has been removed by Man, so that increasingly deep mining is necessary. The heavy cost of working far below the surface, together with the fact that the seams are invariably separated by much useless material, can make coal an expensive product, even though coal-cutting machinery is often used, as in the photograph. This machine is known as a 'coal-plough'; great masses of coal are cut away from the face of the working, to fall on the conveyor-belt in the foreground for rapid transference to underground railroads and so to the foot of the shaft.

9 Mineral oil, like coal, is one of the major fuels of the world; most of the world's vehicles, ships and aircraft are powered by machines (internal combustion, Diesel or jet), which use mineral oil in one form or another. It was probably formed as a result of bacteria acting upon organic matter at the bottom of shallow seas. Once oil has formed it tends to rise under pressure through the pores of the rocks. This is especially common where the rocks have been arched up by pressure to form 'domes' under the surface. The oil may be prevented from rising further by an overlying impermeable layer, which will not permit the oil to pass through. Above the oil usually occurs natural gas, below it a great volume of salt water.

The presence of oil can sometimes be detected by evidence at the surface: by seepage, or by the escape of the natural gas, or by the discovery of deposits of asphalt. But some of the largest oilfields have been found almost by accident in areas where no surface evidence was available. In most cases, following careful investigations by geologists, it is necessary to drill deep bores to see if the oil is actually present. The world's largest petroleum producers are the U.S.A., the U.S.S.R., Venezuela, Kuwait, Arabia, Iran, Iraq and other territories near the Persian Gulf. A small amount of petroleum has been produced in the United Kingdom from oil-shale by exposing the rock to great heat in ovens, but virtually all of the United Kingdom's petroleum requirements are imported.

The photograph below shows part of a large oilfield at Huntington Beach, on the coast of southern California. Note the 'forest' of derricks, each indicating the top of an individual well, even along the very edge of the low cliffs. A network of pipe-lines interlinks the wells and the storage tanks. Oil is pumped for long distances overland through pipe-lines, or

9 Petroleum field in southern California, U.S.A.

is carried across the sea in specially built tankers. The world's largest tanker, recently launched in Japan, can carry 126,000 tons of oil at 16 knots.

10 Many metals occur in the earth's crust in the form of a metalliferous mineral known as an *ore*, where the metal has combined with other elements, so forming sulphides, oxides and carbonates. Iron is a widespread constituent of the earth's crust (probably comprising 5 per cent of the total bulk), but locally it occurs in more concentrated and accessible form, where it can be mined or quarried.

The presence of both iron-ore and coal in the Coal Measures led to the rapid development of the iron and steel industry in many British coalfields during the Industrial Revolution (*c.* 1750–1850). The use of these Coal Measure iron-ores has declined rapidly during the last century and their place in Britain has been taken by ores found in the Jurassic scarplands, extending from Yorkshire to Oxfordshire. The Scunthorpe area, in which the photograph below was taken, is one of the main centres. The iron content of these deposits is low, only about 25 to 30 per cent, but they occur in thick beds near the surface of the ground. They can therefore be cheaply worked by these opencast methods, using large-scale excavators and mechanical shovels, once the overlying bed of soft clays and shales has been removed. On the basis of these extensive deposits a substantial iron and steel industry has developed, notably at Corby. Large amounts of better grade ore are imported from Sweden, Spain, Algeria and elsewhere.

10 Opencast iron-ore mining near Scunthorpe, Lincs.

II. Earth Movements

11 The crust of the earth is continually affected by internal forces, the results ranging from rapid disturbances of the rocks (known as earthquakes) to very slow, large-scale, continent- and mountain-building movements. There are two main groups of forces. One group operates horizontally, so that the crust is crumpled, squeezed up or folded; the other operates vertically, causing uplift or depression of parts of the crust, usually in the form of blocks separated by cracks or *faults*.

This photograph shows the effect of folding on a sedimentary rock, in this case mainly limestone. Denudation by the sea has helped to reveal the crumpled rocks in the form of a section. Although seen here on a small scale, this folding is part of the extensive earth movements, reaching their climax about 35 million years ago, which produced the Alps and other fold-mountain ranges across south-central Europe and on into southern Asia (see Fig. 12).

11 Folded strata at Stair Hole, Lulworth Cove, Dorset

12 An anticlinal ridge in southern Iran

12 In the simplest form of folding, the strata are bent into an upfold (*anticline*) and a downfold (*syncline*). In this photograph an anticline forms a low uniform ridge; a river has cut a gorge right across it, revealing the strata of the left-hand *limb* of the fold, dipping away downwards from the *axis*, or central line of the fold. Many such folds, running broadly parallel to each other, can be seen in southern Persia; these are the Zagros Mountains, part of the great west–east Alpine-Himalayan fold-mountain system which can be traced from western Europe to south-eastern Asia. It is in such folds far beneath the surface that oil and natural gas are commonly preserved (see Fig. 9).

In Great Britain, the London Basin and the South Wales Coalfield are examples of synclinal structures; the Weald of south-eastern England is a complicated example of an anticlinal structure. Simple regular folds are very uncommon; frequently one side of a fold may be steeper than another, and in extreme cases the strata may be forced completely over, to form *overfolds* or *recumbent folds*. Most fold-mountain ranges consist of a complicated series of individual folds, parallel or *en echelon*.

13 Mount Eisenhower, in the Canadian Rockies, near Banff, Alberta

13 As soon as one part of the earth's crust has been elevated, all the forces of earth sculpture seek to wear it down and ultimately to destroy it. An anticline is particularly vulnerable, as rapidly flowing streams can cut gullies in its flanks (as in Fig. 12) and even along its crest. In an area of complicated anticlines and synclines, the former may be worn away so rapidly that in due course the synclines may remain at a higher level, though carved into a series of peaks. This is illustrated in this photograph, where the synclinal arrangement of the rocks can be clearly traced, even at the very summits of the peaks. Mount Eisenhower (9076 feet) is situated to the east of the Bow River valley, swathed with a dense forest of lodgepole-pine in the foreground, between Banff and Lake Louise in the Banff National Park. The mountain consists largely of limestones of mid-Cambrian age; these are well jointed and break away in massive angular blocks, forming a series of steep buttresses. Halfway up the face of the mountain can be seen a prominent terrace, the result of a large band of soft, easily weathered shales. The base of the main cliffs is marked by a distinct change of slope, where the hard limestones rest on much less resistant quartzites. The rocks were folded and uplifted by pressure from the west (left) during the earth movements of early Tertiary times. The main anticline (off the picture on the left) has been largely worn away, leaving the once lower syncline as a striking castellated peak, which was originally known as Castle Mountain. Snowdon, in North Wales, is a further example of a mountain which has a synclinal structure, though now heavily eroded (see Fig. 47).

14 Faulted sandstone rocks, near Bromborough, Cheshire

14 When rocks are subject to stress as the result of earth movements, they may crack or fracture, thus forming either *joints* (see Fig. 2) if there is no actual movement or displacement of the rocks, or *faults*, if the rocks are displaced on either side of the crack. The vertical change of level of the strata may be only a fraction of an inch or thousands of feet, and the rocks may be either raised or lowered. Sometimes a fault is vertical, more usually inclined, as in Fig. 14 above, where several distinct faults interrupt the red and white layers of the Bunter Sandstone. The working of coal mines in Britain and other countries has been greatly complicated by faults which have 'thrown' or displaced seams of coal.

15 A faulting movement usually manifests itself on the surface as an earthquake, whereby vibrations or waves in the rocks travel outwards from the centre of disturbance. The earthquake waves can be recorded thousands of miles away on a delicate instrument known as a *seismograph*. The actual movement may be small, a mere fraction of an inch, while on the other hand the great Alaskan earthquake of 1899 caused a vertical displacement of 47 feet. The effects of an earthquake on the earth's surface vary considerably: landslides occur (see Fig. 52), railway lines and main-pipes are cut, bridges and houses collapse, and in general a wide degree of structural damage may result. This photograph shows how a road has been interrupted where downfaulting caused an extension of the lake. It also reveals that lateral as well as vertical movement is possible along a fault-line; note how the white lines have been displaced 2 feet to the left. This is known as a *tear-fault*, or *strike-slip fault*. The earthquake which disrupted this road occurred on 17 August 1959, in the south-eastern corner of Montana and adjoining parts of Wyoming (see also Figs. 16 and 52).

A most remarkable example of a tear-fault is the 'Great Glen Fault' which runs across the Scottish Highlands from the Moray Firth to the Firth of Lorne. In this area the rocks were torn apart, leaving a gap in which a mass of material sank, in some places a thousand feet, to form the beds of what are now magnificent lochs. Geologists have found evidence to suggest that, at the same time as this trough-fault occurred, the whole of north-eastern Scotland moved about 65 miles along the line of the fault from north-east to south-west. Tear-faults on this scale are, however, uncommon.

Various theories are held about the origins of the so-called Great Rift Valley of East Africa, but there is no doubt that faulting on a gigantic scale has been partly responsible for this remarkable series of features. It can be traced for more than 3000 miles from northern Syria, through the Jordan Valley, the Red Sea, Abyssinia and East Africa (where it divides into eastern and western branches) to the Zambezi River. Another famous fault on a large scale is the San Andreas, which can be traced along the west coast of North America for at least 600 miles, running almost through San Francisco. The earthquake which destroyed that city in 1906 was the result of a horizontal movement of the rocks along this fault of about 20 feet, with a vertical movement of about 3 feet.

15 Road disrupted by the Madison Earthquake, Montana, U.S.A.

16 Fault-scarp, Madison, Montana, U.S.A.

16 The earthquake referred to in Fig. 15 was caused by the creation of two distinct fault-lines, a section of one of which is shown on this photograph. The down-throw to the left, of about 20 feet, produced a low cliff or *fault-scarp* (the term *scarp* refers generally to a steep or near-vertical slope); in the background can be seen the original surface level. This fault-scarp was created by earth movements in a few seconds just before midnight, 17 August 1959. Weathering is already attacking the face of the scarp which, as it is composed of boulder-clay, will soon be smoothed out and will ultimately disappear as a surface feature. The fault continues vertically underground for an unknown distance, and can be traced across country for several miles, like a terrace on the hill-side. Other effects of the earthquake were the cracking of the nearby Hebgen Dam, which fortunately did not collapse, the disrupting of roads (Fig. 15), and the triggering-off of a tremendous rock-slide down the valley-slopes (Fig. 52). Some of the buildings in the nearby Yellowstone National Park were also damaged. Fortunately, this is a thinly inhabited area; if an earthquake shakes the crust beneath a city, destruction and loss of life may be on a large scale, as at Agadir (Morocco) in 1960 and Skopje (Yugoslavia) in 1963.

17 Fault-line scarp, Giggleswick Scar, Yorkshire

17 It commonly happens that faulting brings different kinds of rocks into contact along the line of the fault. In the Craven district of the Pennines a distinct fault-line (known as the *Middle Craven Fault*) can be traced; on the undisturbed higher side the rocks consist of resistant Carboniferous Limestone, but on the down-fault side these rocks lie far beneath the surface, covered by less resistant shales. Denudation has worn away much of the shales, thus causing the line of the fault to stand out as a *fault-line scarp*; that is, a scarp along the fault-line. This differs from the fault-scarp shown on Fig. 16, which is a direct result of faulting; the fault-line scarp is due to faulting followed by prolonged denudation. These cliffs are known as *scars* in Yorkshire. Notice the contrasts between the limestone plateau on the left with much bare rock, the steep prominent face of the scarp, and the soft gentle country of the shales on the right.

Many areas of mountainous country are bounded by fault-line scarps; examples include the Pennine escarpment overlooking the Eden Valley in north-western England, the Cévennes in south-western France, and the vast eastern face of the Sierra Nevada in the U.S.A., rising 8000 feet above the Great Basin at an average angle of about 25°.

III. Vulcanicity

18 Vulcanicity includes all the processes by which solid, liquid or gaseous materials from the interior of the earth are forced into the earth's crust (*intrusion*), or escape on to the surface (*extrusion*). The molten materials injected into the crust, known generally as *magma*, solidify to form intrusive rocks such as granite, gabbro and dolerite, which may later be exposed by denudation.

The molten rock and other materials which reach the surface are called *lava*. When this lava solidifies, it forms extrusive rocks such as basalt, andesite and rhyolite, which can build up land-forms ranging from tiny cones to lofty peaks and vast plateaux.

Molten magma moves towards the earth's surface by forcing its way through lines of weakness in the rocks, such as joints and faults. Sometimes this material solidifies along the bedding-planes parallel to the rock strata to form a *sill*. The photograph shows part of the Great Whin Sill near the coast of Northumberland, consisting of dolerite, or 'whinstone' as

18 The Great Whin Sill at Bamburgh Castle, Northumberland

it is called by the quarrymen. The Sill extends over about 1500 square miles, from the Farne Islands to the western edge of the Pennines, overlooking the Eden Valley. In places the edge of the Sill has been exposed by denudation, forming a striking feature; Hadrian's Wall follows its crest for a number of miles and Bamburgh Castle was built where it forms cliffs near the sea, overlying sandstone of Carboniferous age.

19 Molten material can also force its way vertically *across* the layers of bedded rock, and on solidification it forms a wall of rock termed a *dyke*. Sometimes these dykes occur in large swarms, as in the Isle of Arran, where Fig. 19 was photographed. Here the dykes penetrate the sandstone bedrock, running out from the face of the cliffs across the beaches. A dyke may stand up as a wall or narrow ridge where it has proved more resistant to denudation than the surrounding rocks. On the other hand, as in Fig. 19, it may be softer and so is etched out, in this case by the waves, to form a ditch-like depression.

In addition to sills and dykes, many other land-forms are composed of intrusive rocks. Dartmoor in south-western England is a large

19 Dyke on the coast of Arran, western Scotland

granite intrusion which has been exposed by denudation and worn down to a plateau (see Fig. 2). The general name of *batholith* is given to these intrusions on a large scale, and there are many other varieties, notably *laccoliths*, which arch up the overlying strata.

20 A volcano consists of a vent or opening in the earth's crust, through which various types of material are forced during an eruption, to accumulate as a mountain, more or less conical in shape. Where the cone is built up of successive layers of lava, ash and cinders, it is known as a *composite* volcano, a category which includes most of the loftiest volcanoes in the world.

Around the margins of the Pacific Ocean extends the 'Pacific Ring of Fire', some 25,000 miles in length, and containing about 370 active or recently active, and many thousands of extinct, volcanoes. Part of this ring comprises the Cascade Mountains, extending from California northward for 600 miles through Oregon and Washington to the Canadian frontier. Mount Hood, 11,253 feet high, shown in this photograph, is one of the main Cascade peaks, the highest of which is Mount Rainier (14,408 feet). Though Mount Hood has not had an eruption in historic times, it is not completely extinct, as puffs of steam and sulphurous gas issue from vents near its summit.

20 Mount Hood, in the Cascade Mountains, Oregon, U.S.A.

21 The crater of Ruapehu, North Island, New Zealand

21 Many volcanic peaks culminate in a *crater*, a funnel- or bowl-shaped hollow near the summit, surrounded by an irregular rim. The main pipe, up which passes volcanic material, opens into this crater, which in an active volcano may be filled with pools of hot glowing lava from which issue smoke and steam. In the crater of an extinct or dormant volcano, a lake may accumulate (see Fig. 51).

Ruapehu is a volcano rising to 9715 feet, the highest point in the North Island of New Zealand. The lake in the crater is warmed from below by volcanic activity and so never freezes, in marked contrast to the snow-covered slopes around. The volcano last erupted in 1945, leaving ultimately a crater containing a lake about 260 feet deep. On Christmas Eve, 1953, a natural dam of ash and compact snow resting on the solid lava rim of the crater collapsed, releasing much water which rushed down to the Whangaehu River and, by weakening a bridge, resulted in a rail disaster.

22 The lava produced by a volcanic eruption may be divided into two types according to its mineral composition. 'Acid' lavas have a high melting-point and therefore cool and solidify rapidly after leaving the vent; they accumulate to form high steep-sided cones or domes, such as are found in the *puys* of Auvergne in the Central Massif of France. The 'basic' lavas, on the other hand, with their lower melting-point,

22 The Columbia–Snake lava plateau, U.S.A.

can flow readily for a considerable distance before solidifying. In such a case there is usually no explosive activity, and the lava may quietly well out along a single fissure or a series of parallel fissures. The surrounding landscape may therefore be smothered under successive sheets of this lava, which solidifies as basalt, so forming extensive plateaux.

The photograph above shows one of the largest lava plateaux, that of the Columbia–Snake Basin, which covers an area of more than 100,000 square miles in the north-west of the United States. The earliest flows started about 50 million years ago, and the thickness of each sheet varies from 10 to 200 feet; six or seven individual layers can be distinguished in the photograph. These lavas appear to have 'flooded' the pre-existing landscape, extending into the valleys and in places surrounding peaks which now project through the basalt. The rivers, notably the Snake, have cut down for more than a mile through the basalt, forming some magnificent canyons. During the Ice Age, the Columbia River was diverted further to the south and east by an ice-sheet, and flowed over the cliffs shown in the photograph, known as the 'Dry Falls'. The now abandoned channel to the north, the Grand Coulee, has given its name to a huge dam across the Columbia River, the world's largest concrete structure. The Grand Coulee itself is used as a reservoir for the storage of irrigation water.

A much smaller lava plateau in the British Isles covers part of Antrim in Northern Ireland (see Fig. 3), and another large-scale example can be seen in the north-west of the Deccan in India.

23 One of the most striking examples of solidified acid lava is the Devil's Tower, in the valley of the Belle Fourche River in Wyoming, to the west of the Black Hills of South Dakota. It rises steeply for 865 feet from the scree-slopes around its base, which cover the near-horizontal sandstones, limestones and shales of Triassic and Jurassic age through which the igneous material was forced. The rock which composes the Tower is a greyish phonolite-porphyry, containing prominent crystals of white felspar. On cooling, the magma contracted and formed clear-cut vertical joints, so producing a striking columnar effect; most of these massive columns are five-sided, and are 6–8 feet across. The surface of the Tower, which is almost flat, occupies about an acre; a parachutist who landed here in 1942 found he could not get off, and expert climbers rescued him only after great difficulty.

Its actual origin has caused much argument. Some authorities suggest that it is the surviving neck of a former volcano. Others, more generally supported, believe that it is a large mass of intrusive rock, pushed up like a finger through the sedimentary rocks, and then cooled to its present shape; the Tower was revealed by the denudation of the much softer surrounding rocks. It forms a most spectacular land-form, and was in fact the first National Monument to be designated in the U.S.A.

That spires of lava may be squeezed out of the earth's crust has been proved by the

23 The Devil's Tower, Wyoming, U.S.A.

evidence of many who witnessed the eruption of Mont Pelée in Martinique in the French West Indies in 1902. A *plug* of viscous lava was forced out as a result of the eruption to a height of over 700 feet above the original summit of the mountain. It was, however, largely destroyed by contraction during cooling, and by rapid weathering during the succeeding year. Interesting examples of volcanic rocks in the British Isles are the plugs of Arthur's Seat and Castle Hill in Edinburgh. When a volcano ceases to erupt, lava is left in the crater, and cools to form a mass of solid rock. It may prove so hard that it remains upstanding as a plug or *neck* even when the remainder of the volcano has been destroyed by the weather. Volcanic rock also plays a prominent part in fashioning the scenery of much of the Lake District in north-western England, of Snowdonia in northern Wales and of western Scotland, though there has been no volcanic activity in these areas for a very long period. Snowdon itself (Fig. 47) and Scafell Pike are each largely made of old volcanic rocks.

24 Hot springs, mud-pots, pools of boiling water, fumaroles (emitting steam and gas) and geysers are other minor features of interest found in volcanic regions. Impressive examples of these are found in the Yellowstone National Park in the north-west of the U.S.A., in the North Island of New Zealand, and in Iceland.

A geyser consists of near-boiling water, accompanied by steam, which is emitted with considerable force through a hole in the ground. Some erupt periodically at regular intervals, others more irregularly and spasmodically. The cause is complex, but broadly it is due to heating in the pipe of the geyser; the temperature of the water rapidly increases but convection cannot easily take place, partly because of the length of the pipe, partly because of kinks and twists in it which may allow 'pockets' of steam to accumulate. Ultimately the superheated steam builds up sufficient pressure, expands suddenly, and forces the water in the upper part of the pipe to be violently emitted. Cooler water then flows into the pipe, and the heat increase begins again during the quiescent period.

Old Faithful, one of the most famous geysers in the world, discharges 10,000 to 12,000 gallons of water at each eruption to a height of about 130 feet. The average interval between each eruption is about 63 minutes, though this is quite irregular and varies between 33 and 95 minutes.

There are many hundreds of geysers in Yellowstone. Some are quite small, and erupt regularly every few minutes, while Giant Geyser, which ejected water to a height of 200 feet, has not functioned since 1955. How intimately connected are these various aspects of subterranean activity is shown by the fact that after the earthquake of August 1959 (see Figs. 15, 16) many new geysers started to function, while others became more frequent or powerful, and a few ceased operation.

24 Old Faithful, Yellowstone National Park, Wyoming, U.S.A.

25 Bryce Canyon, Utah, U.S.A.

IV. Weathering and Soils

25 The term *denudation* is used widely to cover all the agencies by which parts of the earth's surface are undergoing destruction, wastage and loss. The loosening, decaying and breaking-up of the rocks are largely due to the various agents of the weather, and are therefore known as *weathering*. Much of this weathered material may move downhill, under the influence of gravity alone; this is known as *mass-movement* or *mass-wasting*. Apart from weathering, earth sculpture is carried out by agents which attack the surface, remove the materials, and deposit them elsewhere; among these agents are streams and rivers, glaciers and ice-sheets, wind and, along the coast, waves, tides and currents. The interrelated processes are referred to as *erosion*, *transport* and *deposition*.

Weathering depends both on the elements of the weather—sun, rain, frost and temperature changes—and on the nature of the rocks: their chemical composition and hardness, whether they have clearly defined joints and bedding planes, and whether they allow water to pass freely through. The weathering processes may be either *physical*, that is, disintegration of the rock into smaller fragments without chemical change, or *chemical*, that is, actual decomposition of some of the mineral constituents of the rock.

The photograph on p. 28 shows the fretted rim of the eastern edge of the Paunsaugunt Plateau in southern Utah, the surface of which lies at about 8300 feet above sea-level. This steep edge is known as Bryce Canyon, though it is not a canyon in the true sense (see Fig. 36), merely a steep-sided nick in the plateau margins. The whole district is now a National Park. A great thickness of brightly coloured limestones, shales and sandstones was laid down in more or less horizontal layers about 55 million years ago, in Eocene times. As these rocks consolidated, vertical cracks and joints developed at right-angles to the bedding planes, forming distinct lines of weakness. Gradually the agents of the weather, helped by running water, have dissected the edge of the plateau, forming the so-called 'Pink Cliffs'. The melting of the snow which lies during the winter for several months to a depth of a yard or more, alternate freezing and thawing, rapid temperature changes during the hot summer, and occasional downpours of rain, all contribute to the decay and removal of the rock. The lines of weakness are emphasised as the bands of softer rock are etched out, leaving the harder formations as pinnacles, columns, bands and ribs; some of these exceed a hundred feet in height. They resemble (in the words of the official *Guide*), '... miniature cities, cathedrals, spires, windowed walls and endless chessmen... challenging the imagination not only with their fantastic forms, but with their color, a riot of pink and red and orange blended with white and gray and cream. Here and there strips of lavender, pale yellow and brown appear—threads of color gone astray from the master design.'

26 Butte with flanking scree-slopes in Arizona, U.S.A.

26 Small flat-topped spurs or *buttes*, projecting from the edge of plateaux or *mesas*, such as that shown above, are familiar land-forms in some of the world's hot deserts. The sandstones and shales along the edge of the plateau crumble slowly but steadily. In dry cloudless regions a very marked diurnal range of temperature occurs, which is caused by powerful heating during the day and rapid cooling at night; the rocks are continuously strained by this successive expansion and contraction, and hence they crack and pieces fall off. The result is to produce a mantle of *scree*, a slope of angular fragments swathing the base of the cliff. The very occasional but heavy rainstorms which are typical of such regions cause torrents to cut gullies into the scree-slopes, which are well shown in the photograph. Note the poor scrubby vegetation, living a slow dormant life until the next rains, and also the horizontal layers of resistant rock which have contributed to the formation of the steep edge of the plateau.

Screes or heaps of broken rock on the slopes of hills and mountains are found not only in desert regions, but to an even greater extent in lands with temperate climates because of the action of frost, which is a powerful disintegrating agent. When the water in the cracks and crevices of rock freezes, it expands and so exerts considerable pressure within the rock mass. This alternate thawing and freezing gradually enlarges the cracks in the rock until pieces break off. The enormous screes to be found on the slopes of some mountains in the English Lake District, which in places extend over 1500 feet in height, are a tribute to this power of frost action. In America scree is known as *talus*.

27 Moist air, though not dry air, has a powerful effect in wearing away the rocks by chemical action. For this reason, in lands with moderate or heavy rainfall the oxygen and especially the carbon dioxide in rain and dew have a very powerful effect, acting as a mild acid which in the course of time may destroy limestone by dissolving the carbonate of lime. Again, granite can be reduced to a mass of clay by decomposition of its felspar minerals (see Fig. 7). In many sandstones the effect is to disintegrate the 'cement' which binds the particles together; this has happened in the case of the Millstone Grit in the photograph below. In this particular example the higher block of rock has proved more resistant to weathering than the rocks below it, thus producing its curious shape.

Weathering caused by moist air or water tends to round off the edges of rocks because these are more exposed. For this reason, hills and mountains in areas of heavy rainfall usually have soft rounded forms, whereas those in dry or semi-arid regions frequently reveal harsh, more angular surfaces (see Figs. 55, 56).

27 Block of Millstone Grit, Brimham Rocks, Yorkshire

28 The effects of gravity on the downward movement of material have been mentioned on p. 29 and again with reference to the formation of scree in Fig. 26. This *mass-wasting* is liable to take place on any steep slope, but it is particularly likely along the coast, where marine erosion can speedily remove material from the base of the slope. Rain-water acts as a lubricant which enables material to slump bodily downwards in a form of land-slip. Along the cliffs at Highcliffe, on the coast of Hampshire, shown in the

28 Mass-movement down the cliff at Highcliffe on the Hampshire coast

photograph (Fig. 28), the crumbling edge of the soft Tertiary sands and gravels has receded inland so rapidly that numerous defensive measures have been taken, including the construction of concrete walls at the base of the cliffs, the encouragement and deliberate planting of vegetation, and the filling of gullies with brushwood (see also Fig. 61).

29 Weathering breaks up the surface rock into progressively smaller pieces. This basic material is colonised by bacteria and the roots of plants, mixed up by the work of earthworms and other animals, and enriched by the decayed remains of plants and organisms to form a soil. These animal and vegetable remains provide humus, which gives the top layers of the soil a darker colour. The photograph opposite shows, below the soil, a layer called the subsoil, where the bedrock is only partly broken up and decomposed, and below this again can be seen the unaltered sandstone. A section such as the one shown in the photograph reveals a series of distinct horizontal layers, often of different colours and textures, which sometimes change sharply, at other times grade into each other. Such a section is termed a *soil profile*, within which each main layer is a *horizon*.

Soils derived from sandstones are often very 'light' and of little value for agriculture because they do not hold water. For this reason the land shown on the photograph is given up to

29 Profile of the soil in an old sand-pit in the New Forest, Hampshire

heathland and sand-pits. Clay on the other hand tends to be very 'heavy', because it holds much water, and so becomes impermeable and sometimes even waterlogged. Loam, a mixture of sand and clay, avoids these extremes and often provides excellent soils. Pure limestone forms no soil at all, since it is almost entirely soluble, but when mixed with clay it may provide good soils called marls. Granite breaks down very slowly, so that most granite areas are covered with only thin and rather poor stony soils. Rocks such as basalt, on the other hand, weather more quickly and produce better soils, often black and rich in minerals.

Of particular interest are the 'young soils', such as the volcanic soils of Java and Italy, and the alluvial soils of the valleys of the Ganges, Nile, Irrawaddy, Hwang-Ho and other great rivers, on which sufficient food is grown to support very dense populations. These are termed 'young' soils because they are comprised of newly formed material, deposited either by the rivers in flood, by volcanic activity, or by other agencies, and have not been subjected to the normal long-period processes of soil formation. Nevertheless, they very quickly form extremely fertile soils, in the case of alluvium because of the fine, well mixed rock material and plant waste from which they are derived, and because of the action of flood-water in depositing annually very fine, well graded silt in level sheets. Moreover, these alluvial plains can easily be irrigated from the rivers. In temperate lands, alluvium also yields fertile soil when it is well drained, as in the polders of the Netherlands and in the Fenlands of England, but it rapidly becomes sour and infertile in badly drained areas. Glacial soils are also examples of young soils which cover much of northern Europe and North America. These soils vary considerably, but commonly they contain a large proportion of stones, even large rocks. The skill of the farmers has converted much of this glacial material into fertile land.

It is evident, therefore, that different kinds of rock may produce different types of soil, though such differences are most marked in 'young' soils and in temperate climates. Over most of the earth's surface, the characteristics of *mature* soils are largely the result of climate, which has a major effect on weathering, on water supply, on the development of humus, and on the vegetation. The world distribution of major soil-types largely follows the pattern of the distribution of vegetation (see the table given below).

Plants obtain their food in liquid form. Water dissolves the soluble minerals in the soil as well as the soluble humus, and carries them downwards; this process is known as *leaching*. Areas of heavy rainfall frequently have soils which have lost much of their mineral content and humus by this leaching; in Malaya, for example, plants do not flourish in open gardens, but must be cultivated in pots or in holes filled with rich soil and a top layer of manure. Conversely, very little leaching takes place in areas of light rainfall, and the soil is able to retain its ingredients in the surface layer. Obviously, where the rainfall is seasonal, leaching occurs only in the wet season. Temperature, as well as rainfall, has a significant effect, for in warm or hot climates evaporation considerably reduces the water content of soils. Moreover, in drier regions evaporation from the top layer causes water and its dissolved mineral salts to be brought up towards the surface by *capillary action*. In cold climates, not only is evaporation ineffective in drying out the soil, but chemical and biological weathering, which produce the humus and the fine-grained particles which help to absorb water, is very limited; hence cold-climate soils, or *podzols*, are usually poor. In hot climates the beneficial effects of high temperatures are often substantially reduced by the rapid rate of leaching in areas of heavy rainfall.

Type of climate	*Vegetation type*	*Major soil type*
Cold climate	Tundra	Tundra soils
Cold climate	Coniferous forest	Podzols
Cool temperate	Deciduous forest	Brown Forest or Brown Earths
Temperate Continental or steppe	Temperate grassland or steppes	Black-earth or Chernozem
Tropical	Savanna or tropical monsoon forest	Tropical Red Earth
Equatorial	Equatorial rain-forest	Laterite

V. Limestone and Chalk Country

30 When water reaches the surface of the earth in the form of rain or snow-melt, part is evaporated, part is absorbed by plants, part runs off down the slopes to join streams which ultimately flow to the sea, and the rest sinks down into the bedrock to form ground-water. Some rocks may be described as *permeable*, and will allow the downward passage of water either because they have a coarse-grained open texture (i.e. they are *porous*) or because they have joints and cracks (i.e. they are *pervious*). Rocks which do not allow any downward passage of water are described as *impermeable*.

Carboniferous Limestone is hard and compact, but it has clear-cut joints and is soluble in water containing carbon dioxide. Thus a stream which rises and flows down over such impermeable rocks as Millstone Grit, and lower down crosses on to limestone, may disappear down a joint, which it will gradually enlarge to form a sink-hole or 'swallow-hole', such as the one shown in the photograph. This particular stream reappears on the surface 350 feet lower

30 Disappearing stream near Malham Tarn, in the Craven District of the West Riding of Yorkshire

down and a mile away, having worked its way underground, now vertically, now horizontally. Thus in areas of limestone a labyrinth of caverns and galleries, underground rivers, lakes and waterfalls may be formed, the haunt of cave explorers. Some underground caves are remarkable in size; the Carlsbad Cavern of New Mexico has a 'big room' which is 300 feet in height, 4000 feet long and 600 feet wide, while a huge cave-system in the Pyrenees is entered by a vertical shaft no less than 1135 feet deep.

The water may finally issue as a powerfully flowing stream near the base of the limestone, where it rests on an impermeable rock. Examples include the stream shown in the photograph, which emerges as the River Aire, the Axe which flows strongly out of the Wookey Hole in the Mendip Hills, and the Peakshole Water issuing from Peak Cavern at Castleton in Derbyshire.

One of the most extensive cave-systems in the world occurs in the plateau of Kentucky in U.S.A. Here there are reputed to be 60,000 'swallow-holes' and many complicated caves. In the famous Mammoth Cave, the surroundings of which form a National Park, there are about 150 miles of mapped caverns, from the lowest of which, 360 feet below the plateau surface, the Echo River emerges; this flows into the Green River, itself a tributary of the Ohio.

31 Carboniferous Limestone is so widespread, and results in such striking landscape features, that it merits a special photograph (opposite). A large part of the Pennines of northern England consists of this rock, forming uplands separated by flat-floored but steep-sided valleys, the famous 'Yorkshire Dales'. These uplands, where they are covered with thin soil, afford grass-covered moorlands, but in places the surface is of bare rock; these 'pavements' are crossed by cracks and furrows which are being enlarged by solution. Harder strata within the limestone produce long grey rock walls or 'scars', sometimes only as high as stairs but sometimes (and particularly where faulting is involved, as in Fig. 17) creating prominent cliff-like features.

The absence of surface drainage, the numerous 'swallow-holes' (Fig. 30), the depressions and hollows produced by the solution of calcium carbonate, the intricate cave-systems, and the dry valleys and gorges are all characteristic of Carboniferous Limestone country. These features can be seen in an emphasised form in the 'Karst' of Yugoslavia, the result of the long aridity of summer and the concentrated though short-lived winter rains. Many parts consist either of bare limestone or are covered with 'garrigue', a thin vegetation of spiny aromatic plants.

The grey Carboniferous or Mountain Limestone of the Pennines is very different from the Jurassic Limestone which stretches across England from Dorset to Yorkshire. The yellow Oolitic Limestones of the Cotswold Hills, for example, are less resistant and produce hill country of lower relief; much of the Cotswolds, because of the deeper soils which cover them, is used for arable and dairy farming.

31 Limestone scenery near Malham, Yorkshire

32 Chalk is a pure form of calcium carbonate, and is highly permeable because of its numerous close joints, thus enabling water to make its way speedily downwards. Solution is also rapid, and it is estimated that each square mile of chalk country in England loses 140 tons of material each year in this way. Surface drainage is confined to a few large streams; occasionally a spring may break out in the floor of a valley, especially after heavy rain.

Chalk forms a long series of hills in south-eastern England, extending from Dorset via Salisbury Plain to the Chiltern Hills and the East Anglian Heights, then appearing further north as the Lincoln and Yorkshire Wolds. The strata are tilted slightly to the south-east or east, and so have a steep slope or scarp to the north-west, with a gentler dip slope to the south-east. The Chalk dips underneath the London Basin, and reappears as the North and South Downs, which are the remains of an anticline formerly covering the whole of the Weald.

Sometimes, where the Chalk is less tilted, it forms more extensive, gently undulating uplands, such as the great expanse of Salisbury Plain. The photograph on p. 38, taken near the southern edge of the Plain, shows many features typical of chalk country—the gently rolling hills, thin soil, and the open expanses of downland covered with short turf. In some areas, notably in the Chilterns and the North Downs, an uneven mantle of clay overlies the Chalk; a large proportion is either covered with woodland (e.g. Burnham Beeches, Box Hill) or is used as arable land.

32 Chalk landscape, near Broad Chalke, Wiltshire

One characteristic feature, clearly visible in the middleground of the picture, is a 'dry valley', a valley (as its name would imply) containing no stream. These often have quite steep walls and abrupt heads, while right-angled bends (as in the photograph) are common. Widely divergent views are held about the origin of such dry valleys. They may have been cut during the last Ice Age, when the Chalk was frozen and impermeable, so allowing surface drainage, while melting snow may have nourished powerful streams. They may possibly have been formed when the rainfall was higher than at present, or the level of water within the rocks (the *water-table*) may have been lowered for some reason, so leaving the valley-floors dry. Whatever the reason, they are striking and impressive.

One interesting man-made feature can be seen in the middleground, a type of terracing known as *lynchets*. These date probably from the Iron Age, and were constructed to provide level strips for cultivation mainly on south-facing slopes.

As chalk is traversed by numerous joints, through which water can easily move underground, it contains a large amount of water, derived from rainfall which has rapidly infiltrated from the surface; it is therefore known as an *aquifer* (literally, water-bearing). This water may appear at the surface as a spring, or it may be tapped by a well, bored down until the water-table is reached. Large amounts of water are obtained from the Chalk for London and other towns in south-eastern England, either from wells or (as in the case, for example, of the city of Portsmouth) from springs.

VI. Rivers

33 A river flows from its source in a spring, a marsh, at the end of a glacier, or simply on a slope where the surface run-off of rain-water or snow-melt collects into a single stream, until it reaches its mouth, which usually opens into the sea. Each river receives tributaries, and so gradually forms a system, occupying a *basin* surrounded by a *watershed.* As rivers flow, they *erode* material from their banks and beds, *transport* it to a lower level, and *deposit* it either on their own valley-floors or on the sea-bottom; in so doing they modify the surface features of their basins. Each section of a river's course reveals characteristic features in the steepness of its gradient and the shape of its valley. As the work of a river goes on steadily and constantly, the idea of a gradual development, or cycle, in age is very apt; it is customary therefore to talk of youth, maturity and old age. The valley of one river may show all three stages: that of youth in its upper course among the mountains, that of maturity in its middle course across the gentler lower country, and that of old age where it wanders across an almost level plain into the sea.

A river is able to carry out its work of erosion in several different, yet related ways. The sheer force of the moving water surging into cracks, with turbulence and eddying, can sweep away material, particularly where the rocks are poorly consolidated. The undercutting of the banks on the outside of a bend can have a powerful effect. A river can carry a considerable load, depending on the volume and the velocity of the water, and on the size and nature of the solid material transported; fine particles are swirled along in suspension, coarser matter proceeds in a series of hops along the bed, while pebbles and even boulders are rolled along when the river is in flood. This load acts as a 'grinding tool', so scouring and excavating the banks and bed. The material itself is progressively worn down, as the fragments are in constant collision both with each other and with the bed of the stream, so that the particles become finer as they move downstream, and of course are easier to transport. In addition, a large amount of matter is borne in solution, notably where a river flows over limestone country.

Fig. 33 on p. 40 shows a 'young' stream, the St Vrain River, which rises on the eastern slopes of Long's Peak in the Rocky Mountains of Colorado, and flows to join the Platte, ultimately via the Missouri to the Mississippi. It is a torrent at this stage, which has cut a steep-sided V-shaped valley in the granite rocks. Its bed consists of pools, pot-holes and boulders, and there are numerous small rapids and waterfalls.

33 A 'young' mountain stream, the St Vrain River, in Rocky Mountain National Park, Colorado, U.S.A.

34 Interlocking spurs, Sail Beck, near Buttermere, in the English Lake District

34 A young stream follows a narrow but winding course, since it tends to flow around obstacles formed by more resistant rocks. These bends are gradually emphasised, since the current tends to be strongest on the outside of the curve, and so the projecting spurs on either side of the river 'interlock' or 'overlap' with each other. The mountain stream shown above, the Sail Beck, flows down from its source high in the Grasmoor massif above Buttermere in Cumberland, and ultimately enters Crummock Water, one of the most attractive of the English Lakes.

Note the barren landscape; among high hills and mountains the strong winds, steep slopes and thin poor soils often prevent the growth of trees and other vegetation, except that which can cling closely to the ground, such as short grasses, bracken, bilberry and heather.

35 Waterfalls provide a most spectacular and attractive feature of a river, particularly in its youthful stage. There are several reasons for their occurrence, but they are basically an interruption of the gradual course of a river by a sudden break of slope. Specific causes include the presence of a bar of hard resistant rock lying across a valley, a sharp well-defined edge to a plateau, a fault-line, and the deepening of a main valley by glaciation, leaving a tributary 'hanging' high above (see p. 58). In many cases, waterfalls and rapids

35 The Yellowstone Falls, Yellowstone National Park, Wyoming, U.S.A.

occur at a point where a resistant rock is succeeded downstream by one less resistant; the greater erosion of the latter may lead only to a steepening of the river bed, in which case rapids will be formed, while a waterfall will develop if the face of the harder rock stands out vertically.

The photograph (Fig. 35) shows the Lower Falls on the Yellowstone River in Yellowstone National Park. In the foreground the river has eroded a deep V-shaped canyon into the yellowish rhyolite rocks which give the Park its name. This gorge has been cut back upstream until the rate of erosion was slowed down by the presence of a mass of much more resistant volcanic rock, so forming a marked vertical step over which the river now falls. From the photograph it is difficult to judge the dimensions of these falls, but they are actually 308 feet in height, or about one and a half times as high as Niagara. Note the great plunge-pool under the fall, into which the water thunders with scouring power.

The waterfall known as High Force in Teesdale has been caused by the river crossing the Great Whin Sill (see pp. 20–1), which has proved more resistant to erosion than the limestone bedrock further down the river. The Niagara Falls have resulted from a horizontal hard bed of dolomite limestone resting upon softer shale; the latter has been worn away more quickly than the limestone, which tends to overhang and form a steep edge over which the water plunges. From time to time, parts of the limestone are broken off, and thus the waterfall moves slowly upstream, leaving a gorge now 7 miles long.

36 If rivers were the only agents in their formation, all valleys would have more or less vertical sides, but in most parts of the world rain, frost and other natural agencies are wearing away the sides of a valley at the same time as the river deepens it, and thus V-shaped valleys are usual. If the rainfall is heavy the V will be wide, if it is light the V will be narrow. Moreover, if the rocks in the valley are resistant, the V will tend to be narrow; if they are soft, the V will be wide. Consequently, a river flowing across very resistant rocks in a dry climate will produce a gorge, especially where it derives its volume mainly from melting snow on lofty mountains far beyond the desert, and is thus able to maintain its flow and its erosive power. One of the best known examples is the Grand Canyon of the River Colorado in Arizona; 300 miles long and about 12 miles wide overall, it has a maximum depth of 6200 feet, cut down through successive layers of limestones, sandstones and shales into the Pre-Cambrian crystalline rocks below.

In some ways the gorge of the Colorado before the river enters the Grand Canyon proper is even more striking. The photograph on p. 44 is taken from the lofty steel bridge which spans the river, carrying the road linking the south and north rims of the Grand Canyon; the difficulty of crossing the river is shown by the fact that while these rims are only 12 miles apart as the crow flies, the distance by road is no less than 215 miles.

The gorge here is 200 feet deep, nearly vertical and in places overhanging. The River Colorado, 50 yards wide at this point, carries an immense load of material, worn by weathering and rain-scour from the valley sides, so that it resembles a stream of thick coffee. This is filling up Lake Mead (the reservoir behind the Hoover Dam downstream from the Grand Canyon) so rapidly that the Glen Canyon Dam farther upstream is at present under

36 The Colorado Gorge, above the Grand Canyon, Arizona, U.S.A.

construction to create a 'settling-reservoir' in which the silt can accumulate. The photograph shows clearly the series of steps by which the sides of the gorge descend to river-level, caused by the alternations of harder and softer rocks down through which the river has cut. Note also the plateau, formed of near-horizontal beds of limestone.

37 This photograph affords another spectacular example of the results of river erosion in a dry climate; this is Rainbow Bridge, made of a vivid orange-coloured sandstone, under which, some 280 feet below, flows Bridge Creek, a tributary of the Colorado. This 'bridge' was formed by a river which wandered over the floor of its valley. Beneath the compact rock of which the 'bridge' is composed lie strata of a much softer pinkish sandstone, into which the river incised itself deeply. Two of its loops approached so closely that ultimately the river cut through the narrow 'neck' and so flowed under the 'bridge' which remained. It has continued to cut down rapidly, and thus the arch has become more and more spectacular, spanning the gorge in one flying leap. Weathering has continued to 'fine down' the 'bridge', and one day in the future it will collapse.

37 Rainbow Bridge, Utah, U.S.A.

It may be mentioned that rock arches can also be formed wholly by weathering in cases where an underlying weaker layer of rock is gradually removed by frost action, temperature change, wind erosion and gravity movement. Examples can be seen in the famous Arches National Monument, also in Utah, and in the 'Angel Window' on the north rim of the Grand Canyon.

38 In its upper course a river mainly wears away its bed because of the speed of its current and the coarse material which it carries; it tends therefore to erode a deep narrow valley. In its middle course the load of the river is composed of less coarse material; the main feature of the river at this stage is the widening of its valley (by *lateral erosion* and *weathering*), while the deepening of the bed (*vertical erosion*) is less marked.

As a river flows round a bend, its current swings powerfully against the outside of the curve, and maximum erosion, even some undercutting, takes place at that point. Thus a river-cliff rises steeply above the outside bank. By contrast, there is little erosion and even some deposition in the slack of the current on the inside of the bend. As the meander develops, swinging laterally across its valley, it may form a level terrace, as shown in the foreground of the photograph below.

Terraces, formed in various ways, may often be observed at different levels on a valley-side. If a river develops a broad flood-plain, and then for some reason resumes its downcutting, remnants of the former surface will survive as

38 River-cliff and terrace on the River Onny, Onibury, Shropshire

terraces above the new channel of the river. This process may be repeated, so that they will occur at different levels.

A large number of river valleys contain terraces consisting of level or gently sloping sheets of gravel, or of finer alluvium, laid over the bedrock. These form well drained, well watered zones above the actual flood-plain of the river, and have frequently provided sites for settlement, roads and railways, and areas of good farming land. The Thames Valley contains three terraces, all of which have played their part in the development of London. Westminster, Battersea and Chelsea have been built on the Flood-Plain Terrace about 20 feet above the river; Holborn and Hyde Park lie on the Taplow Terrace (50 feet), which has also provided much land for cultivation; and the Boyn Hill Terrace (100 feet) includes Pentonville, Clapham and Tooting Commons.

39 Old Man River, seen below, wanders eastwards across the Canadian prairies and ultimately joins the Saskatchewan River. In the middle section of a river's course, lateral erosion is increasingly more effective, and its valley is widened. As the meanders develop, they swing against the valley-sides and form steep-sided bluffs alternately on each bank, as can be seen in the middle of the photograph. The current cuts

39 Old Man River, near Lethbridge, Alberta, Canada

into the upstream sides of these bluffs, and each meander moves gradually downstream, so eroding an increasingly wide valley. In the left and middle foreground of Fig. 39, areas can be seen where the river has deposited part of its load in the slack of the current on the inside of the bends. Note the level surface of the prairies, into which the river is eroding its valley.

40 In its lowest stage, a river is an agent of deposition rather than erosion; it wanders in a series of sweeping meanders over a broad, almost level valley, bounded by low bluffs. Its load consists mainly of finely divided silt and mud, which may be deposited over the valley-floor during times of flood (hence the term *flood-plain*). On such occasions a river may readily abandon part of its course, cutting across the 'neck' of a meander in order to attain a more direct course. The former meander may be left as a 'cut-off' or 'ox-bow' lake. From the air, the pattern of sweeping curves produced by the abandonment of successive meanders can be traced by the deposits of gravel and alluvium; these are shown in very light tones in the photograph below.

40 A winding river in Alaska

41 The Colorado Delta, Gulf of California

41 Deposition of material takes place over the whole lower course of a river, but it is particularly concentrated where the speed of the current is abruptly checked, either where it leaves the mountains and enters a gently sloping plain, or flows into the still waters of a lake, or reaches the sea. A delta is created at the mouth of any river where the deposition of sediment comprising its load exceeds the rate of its removal by tides and currents. Under favourable conditions, a broad sloping bank of sediment is built up very rapidly. As the water becomes shallower, the main channel may split into several branches, known as *distributaries*, along the edge of which mud is deposited, so building out lobes. This photograph was taken on infra-red film, on which water appears in black, thus emphasising the pattern of deltaic deposition and the fan-like distributaries. The Colorado River is at the 'bottom' of the picture and the Gulf of California at the 'top'.

Small deltas can be seen in Britain at points where rivers enter lakes (see Fig. 53), but not where rivers enter the sea. By contrast, in the almost tideless Mediterranean Sea, substantial deltas have been built up at the mouths of the Rivers Nile, Rhône and Po. These facts suggest that a large tidal range prevents the formation of a delta, but this is not the case. The range near the mouth of the Colorado River is high, about 25–30 feet, but a vast amount of sediment is deposited, more than can be removed.

The difficulty is to explain why deltas are not found at the mouths of most rivers; one explanation which has been put forward is that of a general rise of sea-level which has submerged parts or all of some deltas.

42 The terms 'youth', 'maturity' and 'old age' have already been mentioned as symbolising different stages in the development of a river valley. The young river is a powerful agent of erosion, vigorously deepening and enlarging its valley; the river in old age has much less erosive power. Frequently a river may be 'rejuvenated'; that is to say, it regains the erosive power that it formerly possessed as a young river. This may occur as a result either of increased rainfall, or of a fall in sea-level (or an uplift of the land) which can produce a steeper gradient and a more rapid flow.

A rejuvenated river sinks a new channel into the former surface, though maintaining the general pattern of the meanders of its course. These are known as *incised meanders*. Fig. 42 shows the River Wye, which winds and twists through a gorge-like valley, in places forming almost complete loops. Erosion is concentrated on the outside of the curves, where the cliffs are near-vertical, while more gently sloping spurs project from the opposite sides.

If the meander is sufficiently pronounced it will enclose a 'meander-core', as in the centre of the photograph. Sometimes a river may cut right through the 'neck' of the meander and so abandon it, though continuing to erode the 'short cut'. Several large abandoned meanders can be seen in the Wye Valley. Rejuvenation has also caused the formation of a large number of river-terraces (see p. 46).

42 The River Wye between Chepstow and Tintern

43 The River Rhine, flowing between the Hunsruck and the Taunus near St Goarshausen, West Germany

43 The River Rhine flows across the ancient hard rocks of the Middle Rhine Uplands from Bingen to Bonn, creating the famous Rhine Gorge. The Rhine and its tributaries such as the Moselle have divided the uplands into a series of blocks, with a uniformly level surface at about 1500 feet above sea-level, but dropping steeply at the edges for several hundred feet to the valley-floors. The gorge, with castles perched on crags and pleasant small towns at crossing-places, has long been famous in legend and song, and is visited by many tourists. Despite the narrowness of the valley, it is followed by railways and roads, and the river is used by pleasure-steamers and strings of barges towed by powerful tugs. Though many parts of the gorge sides are rocky and precipitous, the south-facing slopes are terraced for vineyards.

The explanation of the relatively flat plateau surfaces cut across by the winding gorges is that the river occupied its present course before the uplift of the plateau in Tertiary times. The ancient rocks had been worn down to a more or less level surface, known as a *peneplain.* When uplift took place, the river continued to cut down and back at virtually the same rate, thus incising its valley deeply; this is known as *antecedent drainage.*

In the English Lake District the drainage is also related to a land surface which has long since disappeared. In this area the old rocks were covered with younger ones, and the whole region was uplifted into an elongated dome in Tertiary times. This cover of younger rocks has long since disappeared (though these can be seen on the margins), but the *radial drainage*, formed when the dome existed, still persists. This is known as *superimposed drainage.*

VII. Glaciation

44 The summit of Mont Blanc (15,782 feet) is a snow-dome, the heart of a complex mountain region in the western Alps; the actual summit lies wholly in France, though the international frontier crosses the massif, and the great face on the right is in Italy. The radiating ridges are mainly of granite; from the snow-fields between them numerous glaciers move downwards and outwards. These parts of the Alps carry a permanent mantle of snow and ice, though the whole Alpine region is snow-covered in winter down to about 3000 feet. The glaciers represent a late stage in the glaciation of the massif, the shrunken remnants of the Ice Age.

44 The Mont Blanc massif, from the south-west

45 Snow, either deposited directly or in the form of avalanches from the surrounding ridges, accumulates in suitable hollows among the mountains to a considerable thickness, and is gradually compacted into masses of white granular ice called *névé* (French) or *Firn* (German). From an upper *névé*-basin, a tongue-like mass of glacier-ice moves out and down along a pre-existing valley, forming a glacier, which ends in a

45 The Gorner Glacier, near Zermatt, Switzerland

'snout'. In the glacier-ice, layers of clear blue or green glassy ice alternate with white granular ice.

A glacier is said to 'flow', as the ice of which it is composed moves downward; the average rate in the Alps is about a foot a day. It is difficult to appreciate the mechanism of glacier-flow—indeed, this is not yet fully understood—but it involves complex physical changes within the mass of the ice. The snout, or lower end of the glacier, indicates the point at which the supply of ice brought down from the high collecting-basins is exactly balanced by the amount of wastage of ice by melting. Fig. 45 shows the end (known as the Boden Glacier) of the Gorner Glacier, which moves in a north-westerly direction, from the complex of snow-fields along the frontier-ridge in the Pennine Alps between Switzerland and Italy, towards the Alpine resort of Zermatt.

The stream in the foreground, after issuing from an ice-cave, wanders over a valley-floor which is strewn with rocks of all sizes left by the glacier, and with sheets of sand and gravel washed out by the melt-water from the load carried by the glacier.

46 The Aletsch Glacier, over 10 miles in length, is the largest in Europe. It originates in a group of extensive snow-fields, at a height of 8–10,000 feet in the heart of the Bernese Oberland, surrounded by fine peaks which rise to 14,000 feet.

As the gradient of a glacier increases, the surface of the ice is split by great cracks or *crevasses*. Some of these run transversely across the glacier, others may lie parallel to the direction of flow. Where the crevasses intersect, the surface of the glacier will form a

46 The Aletsch Glacier, Switzerland

confused labyrinth of 'waves' and pinnacles (*séracs*) of ice separated by deep clefts, as in the foreground of the picture (p. 54). Progress by a climbing party along such a glacier is difficult, and great care is needed.

A glacier is an extremely important agent of erosion, transport and deposition. Frost action causes angular rocks of all sizes to fall on to the surface of the glacier. This debris is slowly carried down with the ice, lying on the surface in a long mound more or less parallel and near to the edge; this is a *lateral moraine*. Where a tributary joins the main ice-stream, two or more lateral moraines may coalesce to form a *medial moraine* down the centre of the main glacier. The photograph shows clearly a large lateral moraine on the far side of the glacier, while several parallel medial moraines faithfully follow the curves of the valley. Some of these are 40 or 50 feet high, consisting of great mounds of rocks of all sizes, while others are just a discontinuous line of isolated blocks.

All this morainic material, together with a vast amount carried frozen within the mass of the glacier, is deposited where the ice finally melts. Towards the snout the surface is debris-covered (see Fig. 45), and in fact it is often difficult to say where the ice actually ends. A crescentic mound of material, a *terminal moraine*, forms around the snout.

47 The Snowdon massif, North Wales

47 The aerial photograph on p. 55 was taken looking almost due south across Snowdon (3560 feet), the highest mountain of southern Britain. The well-known narrow-gauge railway from Llanberis can be traced as it climbs steadily up the spur to the summit-cone, which is crowned by an hotel.

This mountain mass consists mainly of various igneous rocks, known generally as the Snowdon Volcanic Series, of Ordovician age. They are very varied in their resistance to denudation, some forming bold crags and buttresses, others smooth slopes. During a past period of mountain-building, these rocks, and the older slates on which they rest, were folded, and ranges of lofty mountains were formed. Millions of years of denudation wore them down to a flattened dome, part of the surface of which can be seen in the smooth slopes in the middle of the picture.

During the Ice Age it seems that the whole area was at times completely covered by a great ice-sheet, though possibly the higher parts projected above the ice during the later part of the period. Round the flanks small local snowfields accumulated in shallow hollows, which were gradually enlarged by alternate freezing and thawing and by the movement of ice and debris. Gradually the floor of each hollow was deepened and its back wall was steepened to form a basin- or armchair-shaped depression. In these hollows the snow became compacted into *névé* (see Fig. 52), from which small valley-glaciers flowed downwards and outwards. When the ice finally disappeared, a rock-basin remained, known as a *cwm* in Snowdonia, as a *corrie* in Scotland, and generally (after the French name) as a *cirque*. These rock-basins with their steep craggy walls are, as can be seen from the photograph, most striking features of the post-glacial landscape. There are no less than nine of them around Snowdon, five containing lakes. Note how weathering and erosion are still proceeding, cutting gullies, steepening buttresses, and leaving slopes of scree (see p. 30).

When several cirques develop, as in the Snowdon massif, striking relief forms are produced. Steep-sided ridges (*arêtes*), with cols at their lowest points, separate the back-to-back cirques (seen on the left of the photograph). The central surviving mass, Snowdon itself, forms a prominent pyramid-peak (centre background). Probably the most famous example of such a peak, on a much larger scale, is the Matterhorn, 14,713 feet high.

48 When a glacier moves down a former river-valley from the cirque at its head, it powerfully erodes both the sides and floor. Whereas a river normally produces a V profile in the upper part of its valley, a glacier erodes its sides laterally as well as its bed vertically. A glacier erodes in two main ways. The first is by *plucking*, whereby the ice at the base of the glacier actually freezes on to rock-masses, which it tears out. The second is by *abrasion*, in which rocks frozen into the base of the glacier are dragged along, so scraping, polishing and scoring the valley-floor with deep scratches or *striations*. By these means the valley is straightened, projecting spurs are planed off, and the profile becomes that of a broad open U, as shown in Fig. 48. When the ice finally disappears, a river will re-occupy the valley wandering about over the broad, gently sloping floor; this is known as a *misfit*, because the river is so small compared with the size of its valley. Note the pyramid-peak of Cir Mhor at the head of the valley in the photograph.

48 Glen Rosa, Isle of Arran

49 The most striking effects of a glacier on a former river valley are the deepening and widening of its floor and the steepening of its sides. When the ice finally vanished, tributary valleys were often left high above the main valley-floor, and are known as *hanging valleys*. The tributary streams therefore are now compelled to fall abruptly into the main valley as waterfalls, to join the main stream.

The photograph shows the flat floor of Glen Nevis, which curves round the southern and western flanks of Ben Nevis. The main river, the Water of Nevis, wanders around this valley-floor, even dividing (right foreground) to form an island. Near the centre of the picture a tributary spills down from a hanging valley to form a fine waterfall. Note the ice-scraped bare rock on the buttress on the left.

49 A waterfall and hanging valley in Glen Nevis, Scotland

50 Glaciers and ice-sheets are able to transport vast amounts of material, ranging from the finest ground-up rock to enormous boulders; this material is carried on the surface, so forming the moraines shown on Fig. 46, and frozen into the mass of the ice, especially near the base of the glacier. Melt-water streams issuing from the glacier are also powerful transporting agents (see Fig. 45). A vast load is therefore deposited at

50 Boulder-clay, Isle of Arran, Scotland

and beyond the point of melting. Moreover, after the maximum advance of a glacier the edge of the ice decays. Gradually a large area of land is freed from ice, but it remains covered with this glacial material, to which is given the general name of *drift*.

The chief deposit of a glacier is *boulder-clay*, a section of which, cut into by stream erosion, is shown in the photograph above. It consists of a mass of clay, containing stones of all shapes and sizes, varying in character according to their source. Some may have been transported a long distance and are called *erratics*. Sometimes the sheets of clay are more or less evenly spread, sometimes they are gently undulating, sometimes quite hummocky, or the material may lie in long, fairly continuous lines of morainic hills. In places it may comprise swarms of rounded hummocks, varying in size from small mounds to considerable hillocks a mile or more in length and as much as 300 feet high; these are *drumlins*. Boulder-clay covers very large areas of north-western Europe, northern U.S.A. and Canada. In England it lies over much of the lowlands north of a line from the Thames to Bristol.

Boulder-clay is derived mainly from ground moraine. Terminal moraines in many areas are represented by long lines of hills; these are particularly evident in Denmark, East Germany and Poland, where they are known as the Baltic Heights, in places more than a thousand feet high. The melt-water streams at the edge of an ice-sheet sweep out and deposit vast quantities of *fluvio-glacial* material. These deposits include gravels, as in the winding ridges or *eskers* which are common in Sweden and Finland, sheets of sand as in the heath-lands of Germany and Poland, and the pebble-terraces of rivers.

VIII. Lakes

51 A lake is a large hollow in the earth's surface which contains water; its size and permanence depend on the area and depth of the hollow (i.e. its storage capacity), and on the balance between the amount of water received from inflowing streams and from direct rainfall on to the surface, and the loss by outflowing streams, seepage through the floor of the lake, and evaporation. The hollows may be produced by earth movements and volcanic activity (Fig. 51), by erosion (Fig. 53), and by deposition (Fig. 52), but in many cases these different causes may have acted in conjunction with others; thus water can accumulate in an erosion hollow, and be deepened by a natural barrier caused by deposition.

The craters of dormant or extinct volcanoes may provide sites for nearly circular lakes. Crater Lake (below), among the Cascade Mountains in Oregon, U.S.A., has its surface at a height of 6160 feet above sea-level; it is 2000 feet deep, and its jagged rim, 30 miles in circumference, rises in places to over 8000 feet. The water is intensely blue. It is probable that the ancient volcano once exceeded 14,000 feet

51 Crater Lake, Oregon, U.S.A.

in height. The present crater was formed either by a vast explosion which blew off the upper part of the cone or, and this seems more likely, by a collapse of the upper cone combined with subsidence; it is estimated that this took place some 6500 years ago. A new smaller cone was built up by much later volcanic activity within the old crater, and this projects out of the water as Wizard Island; note also its small crater visible in the photograph (see also Fig. 21).

Earth movements have led to the formation of many of the world's most famous lakes. The Dead Sea in Jordan is the deepest part of a depression caused by the sinking of the land between parallel faults; many lakes in the Rift Valley of East Africa have been formed in a similar way. Lough Neagh on the Antrim plateau, Lake Victoria in East Africa, Lake Titicaca on the high Andean plateau in South America, the Caspian Sea and the Great Lakes of North America have all been formed mainly by the bending, sagging or warping of the earth's crust, though various other factors have been involved.

52 An earthquake occurred in south-eastern Montana on 17 August 1959 (see Figs. 15, 16). The shocks caused a vast mass of rock to crash down the steep southern slopes of the canyon of the Madison River, so blocking the valley with a barrier from 200 to 300 feet high. Behind this natural dam the waters of the Madison River were

52 Earthquake Lake, Madison, Montana, U.S.A.

ponded up, and the level rose until a lake 3 miles in length, now officially christened Earthquake Lake, was formed. After some time, the permanence of the dam was threatened by the pressure of the water behind it and, fearful of its collapse with the resulting disastrous floods downstream, engineers cut a relief channel to lower the level; in the photograph, the former higher level of the lake is distinctly indicated by the line of dead trees. The main road along the centre of the valley is now under water, and a new road has been constructed along the northern shore of the new lake.

Many other instances could be given of the formation of lakes by the deposition of material which has subsequently acted as a dam and ponded up water behind it. In 1925 a huge rock-slide dammed the Gros Ventre River in western Wyoming, U.S.A. The following year this natural barrier collapsed, and the resultant flood which surged downstream destroyed the little town of Kelly. Ox-bow lakes (see Fig. 40), the coastal lakes or lagoons of the Baltic Sea enclosed by sand bars, the moraine-dammed lakes of Scotland and the Italian Alps, and the meres formed through the blocking of slowly flowing rivers by sedges, rushes and other aquatic plants (as in north-eastern Belgium), are a few other examples. Another unusual type of barrier-lake is where ice actually forms the dam. Thus the Märjelensee in the Bernese Oberland, Switzerland, lies in the angle between the Aletsch Glacier (Fig. 46) and its rock-walls.

53 The erosive processes of glaciers, discussed on pp. 57–8, can produce depressions of various shapes and sizes wherein water can accumulate, notably in the form of cirques containing small tarns or similar 'basin-lakes' (see pp. 56–7) and in U-shaped valleys in which 'finger-lakes' may lie. Lakes are one of the most characteristic features of a formerly glaciated landscape; much of the charm of the English Lake District, North Wales and the Scottish Highlands is derived from the presence of these sheets of water among the hills and valleys. On the ancient plateaux of Finland and the Canadian Shield are many thousands of irregularly shaped lakes lying in hollows carved out by ice in the crystalline rocks.

The photograph opposite shows the long trough-like Dysynni Valley on the south-eastern flanks of the Cader Idris massif in Merionethshire in Wales. Note the almost flat floor of the glacially eroded valley, with the steeply rising hill-slopes on either side.

A lake is a temporary feature of the earth's surface. For example, the Great Lakes of North America are remnants of enormous lakes which were formed during the Ice Age. Again, the Vale of Pickering and many other valleys in Yorkshire were once filled with water as a result of the damming by ice of the exits to the valleys. Bassenthwaite Lake and Derwentwater in the English Lake District, now separated by a sheet of alluvium 4 miles across, were once a single lake. Fig. 53 shows in the foreground of the picture how material is being built out into the lake, in the form of a small delta, by a stream flowing down from the hills, which may eventually divide the lake into two. The action of rivers, large and small, tends to deposit large amounts of material into lakes, since the speed of a stream, hence its carrying capacity, is abruptly checked when it enters the still waters of a lake. This infilling process has led to the complete disappearance of some lakes.

53 Lake in the Dysynni Valley, Merionethshire, Wales

IX. Deserts

54 About a third of the land-surface of the globe experiences desert or semi-desert conditions, not including the 'cold deserts' or 'ice deserts' of the polar and sub-polar lands (see Fig. 57). In England and other countries with a moderate or heavy rainfall, the surface is covered with grass, trees and other vegetation, and is seldom really dry. This vegetation protects the soil, while the moisture and the roots of plants help to bind the particles together. For these reasons, wind erosion is seldom conspicuous except in sandy areas along the sea coasts, in the heathlands, and in regions of very fine soil such as the Fenlands. In dry regions, however, there is little vegetation to protect the soil and little moisture to bind it together; in these circumstances wind becomes a very powerful agent of transport and erosion.

The final product of desert denudation is sand, which can be moved readily by the wind, to accumulate in the form of low hills or sand-dunes of varying shapes. The dune pattern shown in the photograph below has been formed by a wind blowing constantly from left to right. It is difficult to obtain an impression of scale, but the leeward face of these dunes is

54 Sand-dune patterns in southern Utah, U.S.A.

about 30 feet high. They present a rough crescent shape, with a gentle slope on the windward side and steeper slope to leeward, where eddying by the wind assists in maintaining a slightly concave slope. Dunes rarely occur as individuals, but as an ever-changing 'sand sea' in the form of a whole series of ridges.

Sand-dunes blown by the wind can be very destructive (see p. 76). In humid climates the spread of the sand can be checked by the planting of grasses and pine trees, but in the drier conditions of tropical deserts the problem of preventing the spread of sand to other areas is a much more difficult task. The area covered by a desert is not static; it may increase considerably through the actions of man in the over-grazing or over-cultivation of semi-arid regions, or by neglecting to take action to prevent the spread of sand. On the other hand, a desert may be gradually reduced in size by careful planting of vegetation on its margins, particularly where irrigation water is available.

A desert is often defined as an area with an average rainfall of less than 10 inches. Unfortunately, the rainfall is usually very erratic in areas where the average is low. An average total of 10 inches may mean in practice a series of heavy thunderstorms separated from each other by weeks or even months of absolute drought. In Death Valley, California, there was once a period of three years which was completely rainless.

55 Desert landscape in Jordan

55 The photograph on p. 65 shows a section of rock desert in Jordan, which is being slowly denuded to form a sand desert, as in the portion shown in the foreground. The landscape consists of dissected masses of level-bedded sandstone, with steep angular outlines, rising from a surrounding mantle of rock-waste. The rocky plateau is cut into by weathering, by desert torrents, and by the wind. The last is very effective in emphasising small-scale features of isolated rock masses such as those in the foreground of the picture. The erosive power of the wind involves a triple process. *Deflation* implies the blowing away and removal of any dry unconsolidated material, thus lowering the actual land surface; the material may be carried far away in the form of sand-storms or dust-storms. *Abrasion* is the wearing down of the surface by the load of sand thus carried. Sweeping unimpeded over the open wastes, the wind's powerful 'sand-blast' effect can etch and groove the rocks, picking out the softer strata. The heavier particles are carried near the ground, and so undercutting is marked, producing steep rock faces. A hard stratum may survive for a long time, protecting small isolated pinnacles and 'tables', but ultimately even these are worn away. Note the remains of rock masses to the left of the photograph. This wind-borne material is constantly in a state of movement, impacting one grain against another and against any rock surface. The result is the production of increasingly fine rounded grains; this is known as *attrition*.

56 A striking feature of rock deserts is the steep, craggy, gorge-like valleys, known variously as ravines, *wadis*, gulches and canyons. These are specially well marked when rivers crossing a desert plateau derive their water from far away, possibly from snow-melt on a distant mountain range, as in the case of the River Colorado (Fig. 36). In addition, there is usually some snowfall during the winter, with rapid melting and run-off in spring. But even where rainfall is very infrequent, it is often torrential when it does occur, and these short-lived torrents, known as *flash-floods* in America, can erode rapidly and carry an immense load of material. Fig. 56, of part of the Canyon de Chelly in Arizona, shows the course of a stream, which can be traced from the bleached shingle-banks; when this photograph was taken, the bed was waterless. Great thicknesses of dry sand alternate with patches of deep quicksand; numerous cars driven by incautious visitors along dirt-roads have been bogged down and sometimes abandoned. The steep red walls of the canyon rise for about 500 feet on either side. The higher of the remarkable rock-pillars in the foreground, known as the Spider Rock, is about 800 feet tall. These pinnacles, of brilliant red sandstone of Permian age, have survived because they are defined by clean-cut master-joints, along which weathering and erosion have concentrated. Scrub vegetation is very evident in this photograph, a further indication of near-aridity.

Such plants as the piñon (a small gnarled pine), juniper, sumac, prickly pear and snakeweed manage to survive, the larger trees where their roots can get down to ground-water. In spite of the general aridity, the floor of the canyon is in places the home of Navajo Indians, who grow maize, vegetables and peaches under irrigation, herd sheep, and weave their well-known rugs.

56 The Spider Rock, Canyon de Chelly, Arizona, U.S.A.

57 An ice desert is a convenient term under which to group the great ice-sheets which largely cover the continent of Antarctica (5 million square miles in area) and Greenland (about 600,000 square miles in area). Ice deserts, like hot deserts and rock deserts, are lands in which the climatic conditions are hostile to living things, whether plants, animals or man. These great ice-sheets, in which depths of 7–8000 feet of ice have been found by echo-sounding, are the survivors of the very much larger sheets which once covered much of North America and Europe. Readers of *The Crossing of Antarctica*, by Sir Vivian Fuchs and Sir Edmund Hillary, will know that this type of country is extremely difficult to traverse. Mountain ranges, with peaks projecting above the general level of the ice, steep-sided plateaux, walls and dune-like hills of ice, are common features. The surface varies considerably and is often rough. One of the most dangerous features is the numerous crevasses (clearly visible on Fig. 57), often long and deep, which criss-cross the surface; they are commonly hidden by tenuous snow-bridges that may collapse when weight is placed upon them. The very low temperatures, the frequency of high winds and gales, sometimes accompanied by driving snow and very poor visibility, considerably add to the difficulties of travelling in these regions.

Much of the surface water of the Arctic and Antarctic Oceans surrounding the Polar ice-sheets is frozen permanently or seasonally. During the summer months, blocks of ice (*icebergs*) which have become detached from

57 An 'ice desert' on the margins of Antarctica

the main ice-sheets or from individual glaciers drift equatorwards, thereby menacing shipping. These bergs are exceedingly numerous; probably 10,000 to 15,000 are formed each year in the northern hemisphere alone. The southern hemisphere bergs are sometimes of enormous size; examples have been recorded over 60 miles in length. Because the density of ice (0·9) is only 90 per cent of that of sea-water (1·025), most of the mass of these icebergs floats under water. The 'rule of thumb' that the number of feet of ice projecting above the water equals the number of fathoms below (i.e. 1 to 6) seems by recent investigations to be an exaggeration; proportions of 1 to 4 or 1 to 3 seem to be more likely. Even so, as some bergs tower up for 300 feet or more above the sea, the underwater mass is still enormous. Ice-patrols are maintained north of the shipping lines in the western Atlantic to observe and report upon the movement of the icebergs.

This photograph (Fig. 57), taken by a member of the 1961–3 British Antarctic Survey, shows the edge of a large ice-sheet, reaching the coast to form massive ice-cliffs. Notice the huge crescentic crevasses parallel to the curve of the coast, and also the numerous small bergs and offshore rocky islands.

X. Coastlines

58 Waves are the most efficient agent of marine erosion along the sea coast. A wave exerts a direct effect as it surges against the rocks by compressing the air in cracks, crevices and joints. When it recedes, the compressed air is released with explosive force. This process, repeated infinitely, loosens the surface rock until finally pieces are broken off. In addition, waves carry materials picked up from the beach and use these as very effective battering rams and grinding tools. Hard, solid rocks such as granite are strongly resistant to marine erosion, but nevertheless even these can be eroded over long periods of time. Softer rocks, especially clays, shales, sandstones in which the grains are weakly cemented, and highly soluble limestones, are more easily eroded. For example, the soft glacial sands and clays of East Anglia and Yorkshire are worn away in some places at the rate of 6 or 7 feet a year (Fig. 61). The less resistant rocks may be removed to form bays, whilst the harder rocks stand up boldly as cliffs. In certain circumstances cliffs may be formed even of easily eroded material, such as the 100-foot cliffs of Norfolk composed

58 St Bees Head, Cumberland

of glacial material, or the well-known chalk cliffs of south-eastern England (Fig. 59).

The nature of a cliff depends on a number of factors, such as the height and steepness of the coast, the character of the rocks, their stratification and jointing, and the presence of lines of weakness (such as faults) against which the waves can operate most effectively.

The photograph on p. 70 shows the edge of a flat-surfaced plateau, composed of horizontal strata of New Red Sandstone, which projects boldly into the Irish Sea along the coast of Cumberland. This rock is firm and compact, so forming near-vertical cliffs, but it is soft enough for the waves to undercut it readily at the base, creating numerous caves. Note the fringe of boulders at the foot of the cliff.

59 Although chalk is a soft, easily crumbled rock, it forms very striking cliffs along many parts of the English coast. Waves can cut into these cliffs very rapidly at their base, yet the permeable nature of chalk reduces the amount of water on its surface and thereby greatly reduces the weathering and rounding of its exposed edges. For these reasons, instead of producing a gently sloping surface towards the sea, the chalk stands up boldly as steeply sloping or even vertical cliffs. The undercutting of these chalk cliffs often produces huge rock-falls, and in this way the cliffs recede

59 Chalk cliffs on the Dorset coast, east of Weymouth

inland; since the photograph below was taken, a fall has brought down thousands of tons of chalk to the beach in the extreme right of the picture, though the waves quickly removed most of this debris.

The headland on the left is characterised by clearly marked vertical bedding-planes. These provide lines of weakness which the sea penetrates, so eroding gullies, caves and arches; note Bat's Hole, which completely pierces the headland. This will be enlarged until ultimately the arch will collapse, leaving an isolated pinnacle or *stack*; one can be discerned just to the right of Bat's Hole. Thus the cliff edge is moving steadily landwards, cutting into the chalk downland country. In the centre of the picture, a former dry valley (see Fig. 32), between two parallel chalk ridges, has been cut off abruptly, and is left 'hanging' over the edge of the cliffs.

Within the bay or cove between this headland and another behind the camera, a small gently shelving beach has accumulated. Most of this beach material consists of flints, which occur commonly as bands in chalk. These fragments are pounded by the waves both against the foot of the cliff and against each other, as the uprushing water forces material up the beach, and the backwash drags it back again. Thus the pieces of flint are constantly worn against each other, so forming rounded shingle. Even during calm weather, as in this photograph, this shingle is in constant churning, grinding motion. During storms, the processes are much more active. Sometimes one may visit this section of coast and see huge piles and ridges of shingle; at other times it has been combed down from the beach, and a gently sloping base-platform of bare chalk is exposed.

60 The coast of Pembrokeshire contains some of the most striking examples of coastal scenery in Britain. The rocks, consisting mainly of Old Red Sandstone and Carboniferous Limestone, have been strongly folded and the strata stand up nearly vertically. In the foreground of Fig. 60, in Manorbier Bay, marine erosion of the sandstone strata has produced a remarkable wave-cut rock platform with a pronounced striped effect. This wave-cut platform represents the remains of the cliffs, the edge of which formerly stood much further seaward; as the cliffs recede, the platform grows wider and extends some distance out to sea. The development of this platform slowly reduces the force of the waves by making the sea shallower, particularly as the headlands on either side of the bay also reduce their power; this makes it more difficult for the waves to remove the debris which accumulates at the base of the cliffs. The rate of erosion of the cliffs will therefore be reduced. The bays probably originated because of the presence of groups of small faults, which weakened the rock and facilitated marine erosion. The numerous vertical fissures visible on the headland in the middleground are in some cases also the result of faults or prominent joints, in other cases they are caused by the presence of vertical beds of soft marl among the more resistant sandstones. The main force of the waves is now concentrated on the more exposed headlands, so that in spite of their hard rocks they are being worn slowly back, leaving offshore reefs. Gradually therefore the coast is being cut back, straightened and lowered by marine erosion. A stage may be reached when the coast is stable, that is to say, when it is no longer being worn away by the waves, and when there is a balance between erosion and deposition.

60 Part of the southern Pembrokeshire coast

61 Coastal erosion near Easington, in Holderness, East Riding of Yorkshire

61 The coast of Holderness, as far south as Spurn Head, consists of low yet steep cliffs of boulder-clay, up to 100 feet in height. These cliffs, soft and earthy in character, have receded rapidly inland as a result of marine erosion at their base combined with weathering of the cliff-face itself. It has been estimated that if the present rate of erosion (about 6 to 7 feet a year) has been maintained, a strip of land $2\frac{1}{2}$ miles wide must have been removed since the fifth century. The sites of numerous villages are now well out to sea; many are just names on old maps or in historical records.

This photograph shows the abrupt termination of a road at the cliff edge. This road was used by traffic to the farm on the left until about 1947; indeed, it is shown as a continuous loop of road to the north in the sixth edition of the one-inch O.S. map, though this section had disappeared (replaced by a new inland loop) in the seventh edition, an indication of the speed of marine erosion.

62 The photographs in Figs. 58–61 all show the action of the sea in wearing away the land. Much of the material eroded in this way is moved away and deposited on other parts of the coastline (see Figs. 62–4). The main process by which material is transported is known as *longshore drifting*.

As a wave sweeps material up a beach, the

62 Longshore drifting at Lee-on-the-Solent, Hampshire

backwash drags some of it down again. Should the wave break at an angle (obliquely) on the beach, this material will tend to move up in a similar direction, but the backwash will drag it down the steepest, most direct slope, that is, at right-angles to the shoreline. In this way material will move gradually along the beach. Longshore drifting takes place in a west–east direction along the south coast of England, since the dominant winds and therefore waves are from the south-west. The movement of material from west to east may be halted or reduced by natural features, such as projecting headlands and estuaries, and also by man-made features, such as groynes or moles. At Lee-on-the-Solent (above) and at many other coastal resorts in England and elsewhere, groynes have been built at considerable expense, both to prevent beach material from being transported further along the coast, and also to protect sea-walls and promenades. Such groynes have proved quite successful, though their success has sometimes led to other parts of the coast being starved of material and therefore subjected to erosion on a scale greater than before. Protection of one part of the coast has been achieved only at the expense of other parts. Note from the photograph how the waves tend to sweep the shingle from one side to the other (i.e. from west to east) in the areas between the groynes.

63 Sand-dunes and sand-flats, Goswick, Northumberland

63 The term *beach* is applied to the sloping accumulation of material between the low water spring-tide line and the highest point attained by storm-waves. On a lowland coast a beach may consist simply of an area of sand, sloping gently seawards, vast tracts of which are uncovered at low spring tides. Such beaches are nearly always backed by sand-dunes, because the wind can easily move the loose sand when it dries out at low tide. The prevailing winds tend to move these dunes inland, while others are formed on the upper part of the beach itself; lines of sand-hills may therefore develop over a continuously increasing extent. Large areas on the North Sea coast of the Netherlands, Germany and Denmark have been covered with sand, and much farmland has been thereby destroyed. In the Landes of south-western France, sections of forest have been completely buried in sand, whilst in northern Denmark near Skagen a church-tower stands abandoned among the dunes. Movement of the dunes inland can be limited or checked by the planting of vegetation, such as belts of pine-trees, and, as in this photograph, with marram grass. This plant can readily establish itself in loose sand, which its dense intertwined roots help to bind, while its tufted growth assists in checking the surface-drift of the sand.

64 Various coastal features are produced by the growth of bars and spits of sand and shingle, either across the mouth of a bay or estuary, or at the point where the direction of the coastline suddenly changes, or between the mainland and an island. In these different ways, the sea tends to straighten out or to smooth an irregular coastline.

The photograph in Fig. 64 was taken facing

west. In the top left of the picture is the gently curving coast from Bournemouth—in the centre background—to the sandstone peninsula of Hengistbury Head, which is just off the picture to the left. The middle of the photograph shows the broad shallow bay of Christchurch Harbour, into which flow the Rivers Stour and Avon. Across the mouth of this bay, long narrow sand- and shingle-spits have been built out both from the direction of Hengistbury Head (on the left) and from Mudeford (bottom right), separated by a narrow entrance channel. Old maps and charts show us that the main spit once extended much further eastward (i.e. towards the bottom right corner), parallel to the present coast; probably it was at its longest in about 1880. Sand-dunes have accumulated on both spits, and in recent years numerous holiday bungalows and caravans have been established on them. On several occasions in recent years the main spit has been threatened with breaching by storm-waves. Note how the sand-spits and sandy beaches stand out in white in this photograph; the mud-flats of Christchurch Harbour are a much darker colour.

64 Christchurch Harbour, Hampshire

65 Orford Ness, Suffolk

65 Orford Ness (Fig. 65) is an extensive shingle formation on the coast of Suffolk to the south of Aldeburgh. The right-angled apex in the top right of the picture is Orford Ness itself, which separates the northern part of the spit, shown as a thin white line in the photograph, from the much larger southern part forming the main feature of the photograph. This latter comprises a shingle-spit extending for about 7 miles to its southern tip at North Weir Point (in the foreground of the picture). Beyond this Point, a few shingle patches indicate its line of continuation. The spit was once much longer, but great storms towards the end of last century broke through and destroyed the southern part.

At one time the River Alde reached the sea near Aldeburgh, about 4 miles north of the Ness itself (just off the photograph in the top left). As a shingle-spit grew southwards, the result of longshore drift in that direction, so the River Alde was diverted southwards more or less parallel to the original coastline. In due course the shingle-bar began to develop a slight bend, which steadily became more prominent, so creating Orford Ness. From this the southern portion of the spit continued to grow in a different direction, almost at right-angles to the northern half. Parallel ridges of shingle can be seen on the photograph, indicating the gradual increase in width of the spit. In medieval times the town of Orford, situated in the left centre of the picture, was a port facing the open sea, and its castle, obviously intended to protect a thriving town, was built in 1165; now it is 2 miles as the crow flies from the sea, and 7 miles away up the channel of the Alde. Another river, the Butley, can be seen in the middleground where it enters the Alde, also deflected from its original direct eastward course. Extensive marshes developed inland of the spit, but these for the most part have been reclaimed, and the river is embanked along much of its lower course.

Shingle-spits and ridges are common in many parts of the world; other outstanding examples in England are Chesil Beach in Dorset, Hurst Castle and Calshot Spits in Hampshire, and Dungeness in Kent.

66 When the margins of an upland area are submerged, either because of a rise in sea-level or a fall in the level of the land, an indented coastline is produced, with islands and peninsulas indicating the former uplands, and with inlets of various shapes occupying the former valleys. The most important varieties of this type of coast are (1) a *ria* coast, where submergence affects an area in which hills and river valleys meet the coastline at right-angles; (2) a *fjord* coast, where submergence affects deep glaciated valleys; and (3) a *Dalmatian* or *longitudinal* coast, where the mountains and valleys run parallel to the coastline.

The photograph in Fig. 66 shows part of the north-eastern coast of South Island, New Zealand, in the district known as the Marlborough Sounds, near the town of Picton. An extensive block of upland plateau, the summits of which now rise to 1500–4000 feet above the sea, was heavily dissected by a number of rivers and their tributaries. Submergence of this area took place, partly the result of a tilting of the whole upland block towards the sea, partly because of a general rise of sea-level. The total submergence seems to have averaged about 500 feet, and has produced a most intricate shoreline of a ria-like character. The extent to which the original plateau was dissected is shown by the great number of small inlets (former valleys of tributaries) leading into the large inlets (the former main valleys), and by the close pattern of valleys cutting into the ridges. The main feature of the photograph in the left centre is the inlet known as Tory Channel. Behind is Arapawa Island, rising to ridges of 1500 feet, and behind that again part of Queen Charlotte Sound can be seen.

66 Tory Channel, Marlborough Sounds, South Island, New Zealand

The numerous funnel-shaped inlets separated by promontories, which are shown in the picture, are typical of ria coasts. The south-western coast of Ireland in the counties of Kerry and Cork is another excellent example, as are the coasts of Brittany and of Galieia in Spain.

A further type of indented coast is shown in Fig. 11, where the sea has breached in part the folded and resistant limestone fronting the sea, and has then eroded more swiftly the softer rocks inland to form Lulworth Cove.

67 Milford Sound is an inlet, 12 miles long, on the south-western coast of the South Island of New Zealand. From the water's edge cliffs rise steeply for over 3000 feet to the high plateau surface. The water is deep, depths of 2000 feet being maintained to the head of the inlet. Above the plateau numerous peaks rise to between 5000 and 6700 feet above sea-level. The whole area forms part of Fiordland National Park.

Milford Sound is an example of a fjord, an inlet of the sea which owes its existence to the submergence of a deeply excavated glacial valley. A glacier during a past Ice Age moved down from the mountains, probably along the line of a former river valley, and ultimately produced a steep-sided U-shaped trough (see Fig. 48). A post-glacial rise of sea-level turned this trough into an arm of the sea, but other features of a glaciated valley—truncated spurs and hanging valleys—can be seen. The water is usually shallowest at the entrance, corresponding to the seaward end of a rock

67 Milford Sound, South Island, New Zealand

basin. The steep sides are furrowed by torrents, and there is much bare rock, though coniferous trees grow wherever they can find root-hold.

Fjords are common features on the coasts of Scotland, Norway, British Columbia and the South Island of New Zealand.

Examples of Dalmatian or longitudinal coasts are to be found in Dalmatia itself (the coast of the Adriatic Sea in Yugoslavia), in southern Ireland (Cork Harbour), and along the coast of British Columbia. In these areas, submergence has produced numerous islands, peninsulas and inlets, long and narrow in shape, and in parallel bands, which reflect the trend of the relief—substantially parallel to the coast. Reference to atlas maps showing the coasts of Yugoslavia and of south-western Ireland will quickly show the substantial differences between the Dalmatian and ria types of submerged coasts.

68 In tropical seas where the water temperature never falls below 20° C. (68° F.), colonies of coral polyps can live. These creatures cannot survive for long out of water, and living coral is usually found at or below low-tide level, while conversely they cannot exist at depths much exceeding 25 or 30 fathoms. They require clear oxygenated water, with plentiful supplies of microscopic life as food; they cannot exist in either fresh or heavily silt-laden waters, and hence are never found near river mouths. Food supplies are most plentiful on the seaward side of a reef, so that coral formations tend to grow rapidly outwards. The polyps resemble tiny sea anemones, of many colours and shapes.

As each coral dies, its skeleton accumulates to form white coral limestone.

Gradually, therefore, reefs form around coastlines in tropical seas where these conditions are fulfilled. A *fringing reef* consists simply of an uneven platform of coral closely bordering the coast, with only a narrow shallow lagoon. A *barrier reef* is separated from the mainland by a much deeper, wider lagoon, and an *atoll* consists of a circular, elliptical or horseshoe-shaped reef, enclosing a lagoon, with no central island.

Aitutaki, part of which is shown in this photograph, forms an island of the Lower Cook group, east of the Fiji Islands and 2000 miles north-east of New Zealand. It is particularly interesting in that it exemplifies both fringing and barrier reefs. It is triangular, with sides about 4 miles long; the base of the triangle forms the foreground of the picture. In the north, the background of the photograph, volcanic rock appears on the surface; this is the original foundation of the island, rising steeply from the floor of the Pacific Ocean. On this northern side, it is bordered by a fringing reef. The middleground of the picture shows the lagoon, enclosed by a barrier reef, part of which appears in the foreground, on which stands the little islet of Akaiami. The outer edge of the reef, marked by a line of surf with the open ocean in the foreground, can clearly be seen. As the reef grows, waves wash broken coral on to it, so building up a higher platform. On this accumulates coral sand, well above sea-level, which vegetation, notably coconut palms, helps to fix. Thus a typical 'coral island' develops.

It is not easy to account for the growth of coral islands, particularly barrier reefs and atolls. Even though they rest on some other foundation, these often rise from considerable depths, far below the level at which living coral

68 Aitutaki, Lower Cook Islands, Pacific Ocean

can exist. Various theories have been put forward, usually involving either a gradual sinking of the land, with which the reef-building corals can keep pace, or a general rise of sea-level.

Coral reefs and coral islands are widespread in the Pacific and Indian Oceans and in the Caribbean Sea. The largest example is the Great Barrier Reef off the coast of Queensland, Australia, which is more than a thousand miles long.

69 The mangrove tree and the coral polyp are both inhabitants of the coasts of tropical seas, but where one is found, the other is absent. The coral polyp can only exist in water plentifully supplied with oxygen, and which is clear and free from mud or

69 Mangrove swamps on the coast of British Honduras

silt. The mangrove, on the other hand, is only found where mud is being deposited. It grows on low-lying coasts where the water is calm and protected from surf, but where tidal effects are experienced. The trees form a belt of vegetation progressively extending seawards as the mud-banks spread. On the landward side the mangrove zone gradually changes into fresh-water swamp and, on higher, better-drained ground, into rain-forest.

The term mangrove is given to a number of trees, some twenty or thirty species, not all botanically related, but with certain common characteristics. Their obvious peculiarity is their ability to flourish on these tide-washed mud-flats; at high tide a mass of greyish green foliage appears to float on the water, but at low tide they reveal short stumpy trunks, often supported by a maze of stilt-like aerial roots, or by 'knees' and 'buttresses', through the pores in which the trees can breathe. The swamp at low tide is an area of evil-smelling mud, among which crawl tree-crabs and other creatures.

Mangrove swamps occur notably along the coast of South America near the Amazon and Orinoco Estuaries, along the Caribbean coasts of central America (Fig. 69), along the edge of the Niger Delta, and on the coasts of Sumatra and Borneo. The chief economic value of the mangrove is as a source of tannin, used for converting hides into leather.

XI. Vegetation

The vegetation covering most of the earth has subtle relationships with climate, with the relief of the land, with the soil, and with animals, including man himself. Climate has a major effect, because temperature, rainfall, wind and frost limit the areas within which various plants can grow. Conversely, the vegetation itself has some influence on the climate. The huge forests of the Congo Basin in central Africa, for example, by retaining moisture at the surface of the earth, help to maintain the high humidity of the air. By contrast, the destruction of large areas of forest in many parts of the U.S.A. has tended to reduce the humidity and probably the rainfall of these areas, and this has made it more difficult to re-establish forests.

The relief of the land has a major effect upon the vegetation. In areas of moderate or even heavy rainfall the forests which form the natural vegetation of the lowlands are replaced by grasslands in mountainous regions and, at still higher levels, by tundra and bare rock. This is because of the decrease in temperature with height, and because the stronger winds cause intense evaporation. By contrast, in drier lands, as for example in many parts of East Africa, the grassland or 'parkland' of low-lying areas may change to zones of forest at higher altitudes because of the appreciable increase in rainfall.

Soils are closely related to the plants which live in them. Very sandy soils, for example, cannot easily hold moisture, and only certain small plants, such as ling, gorse and heather, are able to adapt themselves to the conditions; these plants produce only a thin layer of humus and thereby do not much help its water-holding properties.

Man and his animals have been profoundly affected by the natural vegetation, and in their turn have effected substantial changes in this vegetation cover. Early Man in parts of Europe avoided the lowlands and settled on the higher, better-drained lands; these soils were easier to clear and to cultivate. Modern Man has removed much of the natural forest and even more of the natural grasslands, and has replaced the vegetation with plants of value to him. Much of the land now classified as savanna in Africa may once have been covered with forest; centuries of shifting cultivation by the African peoples and the grazing of their animals have prevented the replacement of the former forest cover. But huge areas of tropical forests, in the Congo and Amazon Basins and elsewhere, are still formidable obstacles to settlement.

70 In a broad sense, rainfall is the most significant factor in determining the type of vegetation. Scrub and thornbushes, separated by areas of bare sand and rock, characterise areas where the rainfall is deficient (see Figs. 26, 55, 56). Grassland or 'parkland' is general where the rainfall is somewhat greater, and forest occurs in regions of abundant rainfall. The type of forest—cool temperate forest, warm temperate forest, tropical forest—is largely the product of differing conditions of temperature.

The photograph in Fig. 70 was taken at the edge of the Equatorial Forest (or *Tropical Rain-forest*) in the west of the Congo Basin, where it extends over an area equal in size to western Europe. The hot, wet climate provides conditions for the rapid growth of luxuriant vegetation. The 'roof' of this forest is formed by hundreds of different kinds of tree up to 150 feet tall, which carry most of their branches and leaves high up to reach the light, and so create an almost continuous layer or 'canopy' of vegetation. This picture shows mainly the

70 Tropical Rain-forest in the Congo Basin

smaller trees, the giant ferns, the thick woody climbers (*lianas*), and various plants which grow on each other (*epiphytes*), occurring at an intermediate level. Ferns and large fleshy plants flourish nearer the ground, thriving on the mass of decayed vegetation covering the soil. The canopy is more clearly seen from the air, as in Fig. 71.

The various trees and plants produce leaves, flowers and fruit at very different times of year. At any time, any plant, or even a single branch of a plant, may be in leaf, flower or fruit. The forests therefore give an impression of a monotonous uniform luxuriant greenness throughout the year; hence although they shed their leaves at some time or another, they are commonly called 'Evergreen'. A remarkable fact about these forests is the enormous number of different species of trees; it has been estimated that there are 30,000 in Indonesia, compared with 2000 in the British Isles. Despite the difficulty of exploiting these forests, the extraction and export of hardwoods such as mahogany, rosewood, ebony and greenheart is a substantial industry in many areas. There are many trees, both hard- and soft-woods, sometimes known only by their native or botanical names; some of these are now used commercially.

The distribution of Equatorial Rain-forest is limited strictly to the Congo and Amazon Basins, the coasts of Central America and West Africa, Malaya, Indonesia and the lowlands of southern Vietnam and Cambodia. Even in these areas, much has been destroyed and replaced by 'secondary forest', which has a far denser undergrowth, contains a large number of small fast-growing soft-woods, and is difficult to penetrate.

71 Swamp-forest and Rain-forest in the Niger Delta

71 The swamp-forest associated with the deltas of rivers such as the Niger, and the mangrove forests (Fig. 69) which form so prominent a feature of shallow coasts and river mouths, are both different in kind from the true rain-forest; they form a strange mixture of forest, shrub, dense undergrowth, swamps, lakes and rivers. In the Irrawaddy Delta, by contrast, much of the original forest has been cleared for rice cultivation and the forest is now restricted to areas where the flood water is difficult to control.

72 Tropical Monsoon Forest on the Burma–Thailand border

72 Most regions on or near the Equator have short periods of dry weather, but further north and south the dry and wet seasons are much more pronounced, as a result of which most of the trees lose their leaves during the dry season. The most widespread type is called *Tropical Monsoon Forest*. These forests are generally mixed in character, but less so than the Equatorial Forests, and almost pure stands of a single species occur, as in the teak forests of Burma. The dry period is used as a resting season by the trees, for which reason they are smaller and have thicker barks than those of the rain-forests and are less luxuriant in form. The photograph below shows monsoon forest on hills near the Burma–Thailand border. In much of the isolated hill country of Burma, Thailand and the Malay Peninsula large areas of this monsoon forest remain, though near centres of population most of the original tree-cover has long been cleared. This has been replaced by rice cultivation in the lowlands and by rubber plantations on lower slopes, while a secondary forest of dense tangled woodland covers much of the hills.

73 Forests, similar in type to the Tropical Monsoon Forest, form the characteristic vegetation of most tropical regions which have both a heavy total rainfall and a pronounced dry season. In the drier areas, forest is replaced by grassland. This *Savanna* or *Tropical Grassland* varies considerably, from the 'parklands' which border the forested areas, to scrub and coarse grassland nearer the margins of semi-arid regions. Commonly it consists of an uneven cover of tall grass, growing rapidly during the wet season to a height of 6 to 12 feet as a stiff yellow straw-grass crowned with silvery spikes. Clumps of trees, wedge- or umbrella-shaped as a result of the frequent strong winds, and resistant to long periods of drought, grow in hollows where ground-water approaches the surface; these include palms, baobabs, acacias and a variety of thorny bushes. As the length of the winter dry season increases and the total rainfall decreases, the trees gradually disappear from the landscape, the grass becomes shorter and sparser, patches of bare sand are more frequent, and the thorny shrubs appear more stunted. Here the savanna degenerates into the scrub vegetation of the semi-deserts. The photograph below shows a typical savanna landscape in Southern Rhodesia; note the low scattered trees, the tall grass which limits the view, and the numerous bushes.

73 Savanna lands in Southern Rhodesia

74 Coniferous Forests in the Laurentian Shield, Canada

74 Areas of warm and cool temperate climates are characterised by markedly different kinds of forest. In the warm temperate lowlands in the south-east of the U.S.A., in the southern parts of Brazil, China and Japan and in eastern Australia, temperatures are high in summer and the rainfall is plentiful and well distributed. In these conditions, a type of forest has developed similar in some ways to the Equatorial Rain-forest. Epiphytes, lianas, mosses and lichens are common; there is a great variety of trees, from bamboos, palms, tree ferns and magnolias to cypresses, evergreen oaks and pines. But the trees are largely of uniform height and smaller than those in equatorial regions; the undergrowth is usually thick and quite difficult to penetrate.

In warm temperate lands with a dry summer season, the so-called 'Mediterranean' lands, the natural vegetation of open oak woodlands has largely been destroyed by Man to make room for the vine, olive, citrus fruits and other cultivated crops, and by fire, accidental or deliberate. In areas which are too rough to be cultivated, the former oak woodlands have been replaced by *maquis*, comprising dense thickets of creepers and evergreen shrubs such as oleander, heath, myrtle and rosemary. Many of these shrubs have small hard leaves covered with hair or gum to reduce transpiration in the hot dry summer season.

The natural vegetation of much of central and western Europe and eastern North America south of the St Lawrence River consisted of *Cool Temperate Deciduous Forest*; most of this has long since been destroyed by Man, but it probably contained extensive areas of beech, oak and willow, as well as mixed woodlands of poplar, elm, sycamore, chestnut and (in North America) maple, hickory, cedar and spruce. North of these deciduous forests, in a great belt stretching from Siberia to Scandinavia and from the St Lawrence River to Alaska, are the *Coniferous Forests*, a photograph of which is given in Fig. 74. The picture, which was taken near Goose Bay, Labrador, shows the region of the Hamilton River; it gives an impression of the vast area covered by this type of forest, the monotony of the vegetation caused by the very small number of species, and their similarity in shape and size. Only a few types of tree, mainly larch, spruce, pine and fir, can withstand the long cold winters, when snow covers the surface for many months. On the very thin soils of much of the Canadian Shield and of the high plateaux of Scandinavia, and in the bleaker Tundra climate still further north, the trees become appreciably smaller and are of value only as firewood and pulpwood. The best stands are found in the south of the climatic zone and near the coasts. It has been found necessary to re-afforest large areas in order to provide a continuous supply of timber accessible to man.

75 Cool Temperate Grasslands are found in the interiors of continents in temperate latitudes, where the rainfall is inadequate for forest growth but sufficient for smaller plants, of which the various grasses are typical. The most extensive examples are the *steppes* of Eurasia, the *prairies* of North America, the *pampas* of the Argentine, the *downs* of south-eastern Australia and the *veld* of South Africa; a photograph of the last, in the Orange Free State, is shown below. The monotonous aspect of these vast, gently undulating, treeless plains, which appear to stretch endlessly beyond the horizon, is well illustrated in the picture. The light cumulus clouds and considerable sunshine are typical of these dry lands in summer, though in winter the weather is often severe, especially in the prairies of Canada.

75 Cool Temperate Grassland in the Orange Free State, Republic of South Africa